Glencoe McGraw-Hill

Math Connects
Course 2

Study Guide and Intervention
and Practice Workbook

McGraw Hill Glencoe

To the Student This *Study Guide and Intervention and Practice Workbook* gives you additional examples and problems for the concept exercises in each lesson. The exercises are designed to aid your study of mathematics by reinforcing important mathematical skills needed to succeed in the everyday world. The materials are organized by chapter and lesson, with one *Study Guide and Intervention and Practice* worksheet for every lesson in *Glencoe Math Connects, Course 2*.

Always keep your workbook handy. Along with your textbook, daily homework, and class notes, the completed *Study Guide and Intervention and Practice Workbook* can help you review for quizzes and tests.

To the Teacher These worksheets are the same as those found in the Chapter Resource Masters for *Glencoe Math Connects, Course 2*. The answers to these worksheets are available at the end of each Chapter Resource Masters booklet as well as in your Teacher Wraparound Edition interleaf pages.

The McGraw·Hill Companies

Send all inquiries to:
Glencoe/McGraw-Hill
8787 Orion Place
Columbus, OH 43240

ISBN: 978-0-07-881054-1
MHID: 0-07-881054-X

Study Guide and Intervention and Practice, Course 2

Printed in the United States of America
6 7 8 9 10 MAL 14 13 12 11 10

CONTENTS

Lesson/Title		Page

1-1 Study Guide and Intervention

A Plan for Problem Solving

Four-Step Problem-Solving Plan

When solving problems, it is helpful to have an organized plan to solve the problem. The following four steps can be used to solve any math problem.

1. **Understand** – Get a general understanding of the problem. What information is given?

2. **Plan** – Select a strategy to solve the problem and estimate the answer.

3. **Solve** – Carry out your plan to solve the problem.

4. **Check** – Determine the reasonableness of your answer compared to your estimate.

Example 1 Use the four-step plan to solve the problem.

RECREATION A canoe rental store along the Illinois River in Oklahoma has 30 canoes that it rents on a daily basis during the summer season. If canoes rent for $15 per day, how much money can the store collect for canoe rentals during the month of July?

Understand	You know that they rent 30 canoes per day for $15 each. You need to determine the total amount of money that can be collected during the month of July.
Plan	First, find the total amount of money that can be collected each day by finding the product of 30 and 15. Next, multiply the previous result by 31, the number of days in July. You can estimate this result by 30. $30 \times 15 \times 30 = 13,500$
Solve	Since $30 \times \$15 = \$450$, the canoe rental store can collect $450 in rental fees each day. This means the total amount of money that could be collected during the month of July is 450×31 or $13,950.
Check	Is your answer reasonable? The answer is close to the estimate of $13,500.

Exercises

Use the four-step plan to solve each problem.

1. **MONEY** Colin works for his dad during summer vacation. His dad pays him $5.20 per hour and he works 20 hours per week. How much will Colin earn during his 8-week summer vacation?

2. **BOOKS** A student assistant in the school library is asked to shelve 33 books. If he puts away 9 books the first hour and then 6 books each hour after that, how long will it take him to shelve all 33 books?

3. **SHOPPING** Alicia bought a $48 sweater on sale for $25 and a $36 purse on sale for $22. How much did Alicia save?

4. **MAIL** It cost Ramon $3.73 to mail a package to his grandmother. The post office charged $2.38 for the first pound and 45 cents for each additional pound. How much did the package weigh?

1-1 **Practice**

A Plan for Problem Solving

Use the four-step plan to solve each problem.

1. **ENGINES** A car engine turns 900 revolutions per minute while idling. How many revolutions does a car engine turn in one second while idling?

2. **DISTANCE** While traveling in Montana from Butte to Sidney, Mr. Kowalski, recorded that the distance from Butte to Sidney was about 6 times the distance from Butte to Bozeman. Bozeman lies between Butte and Sidney. If the distance from Butte to Bozeman is 82 miles, what is the approximate distance from Bozeman to Sidney?

3. **NUMBERS** What are the next two numbers in the pattern?
   3.1, 3.11, 33.11, 33.111, _____ , _____

4. **TIDES** The Bay of Fundy in Nova Scotia, Canada is known for large tides. On a particular day low tide was at 2.3 feet. The tide then rose 6.6 feet every hour for the next six hours. What was the height of high tide on that particular day?

5. **BASKETBALL** If team A won by 2 points what was the number of points scored by team A in the 3rd quarter?

Team	Quarter Scores 1st 2nd 3rd 4th				Final Score
A	21	18	?	17	?
B	15	19	20	25	79

6. **COOKING** A cake recipe requires a total 16 tablespoons of butter for one cake, some for the batter and some for the frosting. If 4 tablespoons of butter are needed for the batter for one cake, how many tablespoons of butter are needed for the frosting if Samantha wants to bake three cakes?

1-2 Study Guide and Intervention

Powers and Exponents

$$\overset{\text{Exponent}}{3^4} = \underbrace{3 \cdot 3 \cdot 3 \cdot 3}_{\text{common factors}} = 81$$

Base

The **exponent** tells you how many times the **base** is used as a factor.

Example 1 Write 6^3 as a product of the same factor.

The base is 6. The exponent 3 means that 6 is used as a factor 3 times.
$6^3 = 6 \cdot 6 \cdot 6$

Example 2 Evaluate 5^4.

$5^4 = 5 \cdot 5 \cdot 5 \cdot 5$
$\quad = 625$

Example 3 Write $4 \cdot 4 \cdot 4 \cdot 4 \cdot 4$ in exponential form.

The base is 4. It is used as a factor 5 times, so the exponent is 5.
$4 \cdot 4 \cdot 4 \cdot 4 \cdot 4 = 4^5$

Exercises

Write each power as a product of the same factor.

1. 7^3
2. 2^7
3. 9^2
4. 15^4

Evaluate each expression.

5. 3^5
6. 7^3
7. 8^4
8. 5^3

Write each product in exponential form.

9. $2 \cdot 2 \cdot 2 \cdot 2$

10. $7 \cdot 7 \cdot 7 \cdot 7 \cdot 7 \cdot 7$

11. $10 \cdot 10 \cdot 10$

12. $9 \cdot 9 \cdot 9 \cdot 9 \cdot 9$

13. $12 \cdot 12 \cdot 12$

14. $5 \cdot 5 \cdot 5 \cdot 5$

15. $6 \cdot 6 \cdot 6 \cdot 6 \cdot 6$

16. $1 \cdot 1 \cdot 1 \cdot 1 \cdot 1 \cdot 1 \cdot 1 \cdot 1$

1-2 Practice

Powers and Exponents

Write each power as a product of the same factor.

1. 5^7

2. 2^4

3. 7^2

4. 10^5

5. 3^3

6. 6^8

7. *four to the eighth power*

8. *eight cubed*

9. *ten squared*

Write each product in exponential form.

10. $9 \cdot 9 \cdot 9 \cdot 9 \cdot 9 \cdot 9$

11. $1 \cdot 1 \cdot 1 \cdot 1 \cdot 1$

12. $2 \cdot 2 \cdot 2 \cdot 2 \cdot 2 \cdot 2 \cdot 2$

13. $6 \cdot 6 \cdot 6 \cdot 6 \cdot 6 \cdot 6 \cdot 6 \cdot 6 \cdot 6$

14. $5 \cdot 5$

15. $4 \cdot 4 \cdot 3 \cdot 3 \cdot 3 \cdot 3 \cdot 3 \cdot 3$

Evaluate each expression.

16. 4^3

17. 1^{11}

18. 2^5

19. 10^3

20. 9^3

21. 8^1

22. *five to fourth power*

23. *7 squared*

24. *zero to the sixth power*

Use a calculator to determine whether each sentence is *true* or *false*.

25. $2^8 = 8^2$

26. $17^2 < 172$

27. $3^2 > 1^{19}$

Order the following powers from least to greatest.

28. $7^2, 5^3, 3^4, 2^5$

29. $4^3, 1^{13}, 12^2, 8^3$

30. $3^9, 5^7, 7^5, 9^3$

31. **INTERACTIVE MAPS** Mansi is using an interactive map on her computer that allows her to zoom in or zoom out. Each time she zooms out the scale of the map increases by a power of ten. If she zooms out four times the scale is 10^4 times greater. Write this number in standard form.

32. **BACTERIA** A lab technician observed 5 bacteria growing in a lab dish. One hour later he observed 25 bacteria. Every hour he notices about 5 times as many as the hour before. After several hours of observation, he determined the lab dish had 5^9 bacteria. Use a calculator to find the number in standard form that represents the bacteria in the lab dish.

1-3

Study Guide and Intervention

Squares and Square Roots

The product of a number and itself is the **square** of the number. Numbers like 4, 25, and 2.25 are called **perfect squares** because they are squares of rational numbers. The factors multiplied to form perfect squares are called **square roots**. Both $5 \cdot 5$ and $(-5)(-5)$ equal 25. So, 25 has two square roots, 5 and -5. A **radical sign**, $\sqrt{}$, is the symbol used to indicate the *positive* square root of a number. So, $\sqrt{25} = 5$.

Examples

1 Find the square of 5.

$5 \cdot 5 = 25$

2 Find the square of 16.

16 $\boxed{x^2}$ $\boxed{\stackrel{\text{ENTER}}{=}}$ 256

3 Find $\sqrt{49}$.

$7 \cdot 7 = 49$, so $\sqrt{49} = 7$.

4 Find $\sqrt{169}$.

$\boxed{\text{2nd}}$ $[\sqrt{}]$ 169 $\boxed{\stackrel{\text{ENTER}}{=}}$ 13

So, $\sqrt{169} = 13$.

Example 5 A square tile has an area of 144 square inches. What are the dimensions of the tile?

$\boxed{\text{2nd}}$ $[\sqrt{}]$ 144 $\boxed{\stackrel{\text{ENTER}}{=}}$ 12   Find the square root of 144.

So, the tile measures 12 inches by 12 inches.

Exercises

Find the square of each number.

1. 2

2. 9

3. 14

4. 15

5. 21

6. 45

Find each square root.

7. $\sqrt{16}$

8. $\sqrt{36}$

9. $\sqrt{256}$

10. $\sqrt{1,024}$

11. $\sqrt{361}$

12. $\sqrt{484}$

Lesson 1-3

1-3 Practice

Squares and Square Roots

Find the square of each number.

1. 2

2. 8

3. 10

4. 11

5. 15

6. 25

7. What is the square of 5?

8. Find the square of 16.

9. Find the square of 21.

Find each square root.

10. $\sqrt{64}$

11. $\sqrt{121}$

12. $\sqrt{169}$

13. $\sqrt{0}$

14. $\sqrt{81}$

15. $\sqrt{289}$

16. $\sqrt{900}$

17. $\sqrt{1}$

18. $\sqrt{484}$

PACKAGING An electronics company uses three different sizes of square labels to ship products to customers. The area of each type of label is shown in the table.

Labels	
Type	**Area**
Priority:	100 cm^2
Caution:	225 cm^2
Address:	144 cm^2

19. If the length of a side of a square is the square root of the area, what is the length of a side for each label?

20. How much larger is the Caution label than the Address label?

21. **RECREATION** A square hot tub is outlined by a 2-foot wide tile border. In an overhead view, the area of the hot tub and the border together is 144 square feet. What is the length of one side of the hot tub itself?

1-4 Study Guide and Intervention
Order of Operations

Use the **order of operations** to evaluate numerical expressions.

1. Evaluate the expressions inside grouping symbols.
2. Evaluate all powers.
3. Multiply and divide in order from left to right.
4. Add and subtract in order from left to right.

Example 1 Evaluate $(10 - 2) - 4 \cdot 2$.

$(10 - 2) - 4 \cdot 2 = 8 - 4 \cdot 2$ Subtract first since $10 - 2$ is in parentheses.
$\quad\quad\quad\quad\quad = 8 - 8$ Multiply 4 and 2.
$\quad\quad\quad\quad\quad = 0$ Subtract 8 from 8.

Example 2 Evaluate $8 + (1 + 5)^2 \div 4$.

$8 + (1 + 5)^2 \div 4 = 8 + 6^2 \div 4$ First, add 1 and 5 inside the parentheses.
$\quad\quad\quad\quad\quad\quad = 8 + 36 \div 4$ Find the value of 6^2.
$\quad\quad\quad\quad\quad\quad = 8 + 9$ Divide 36 by 4.
$\quad\quad\quad\quad\quad\quad = 17$ Add 8 and 9.

Exercises

Evaluate each expression.

1. $(1 + 7) \times 3$
2. $28 - 4 \cdot 7$
3. $5 + 4 \cdot 3$

4. $(40 \div 5) - 7 + 2$
5. $35 \div 7(2)$
6. 3×10^3

7. $45 \div 5 + 36 \div 4$
8. $42 \div 6 \times 2 - 9$
9. $2 \times 8 - 3^2 + 2$

10. $5 \times 2^2 + 32 \div 8$
11. $3 \times 6 - (9 - 8)^3$
12. 3.5×10^2

Lesson 1-4

1-4 Practice

Order of Operations

Evaluate each expression.

1. $(2 + 9) \times 4$

2. $8 - (5 + 2)$

3. $(15 \div 3) + 7$

4. $(14 + 7) \div 7$

5. $5 \cdot 6 - 12 \div 4$

6. $8 \div 2 + 8 - 2$

7. $16 - 8 \div 2 + 5$

8. $15 - 3 \cdot 5 + 7$

9. 7×10^3

10. $2 \times 5^2 + 6$

11. $7 \cdot 2^3 - 9$

12. $27 \div 3 \times 2 + 4^2$

13. $6^3 - 12 \times 4 \cdot 3$

14. $(15 - 3) \div (8 + 4)$

15. $(9 - 4) \cdot (7 - 7)$

16. $8 + 3(5 + 2) - 7 \cdot 2$

17. $5(6 - 1) - 4 \cdot 6 \div 3$

18. $(5 + 7)^2 \div 12$

19. $12 \div (8 - 6)^2$

20. $(7 + 2)^2 \div 3^2$

21. $(11 - 9)^2 \cdot (8 - 5)^2$

22. $64 \div 8 - 3(4 - 3) + 2$

23. $8 \times 5.1 - (4.1 + 1.4) + 7.1$

For Exercises 24 and 25, write an expression for each situation. Then evaluate the expression to find the solution.

24. **LAWN AREA** The Solomons need to find the area of their front and side yards since they want to reseed the lawn. Both side yards measure 3 meters by 10 meters, while the front yard is a square with a side of 9 meters. They do not need to reseed a portion of the front yard covering 16 square meters where a flower bed is located. What is the area of the yard that the Solomons want to reseed?

25. **COMMUNITY SERVICE** Jariah volunteers at the hospital during the week. She volunteers 3 hours on Monday and Thursday, 4 hours on Saturday and Sunday, and 2 hours on Tuesday. How many hours does Jariah volunteer at the hospital during the week?

1-5

Study Guide and Intervention

Problem-Solving Investigation: Guess and Check

When solving problems, one strategy that is helpful to use is guess and check. Based on the information in the problem, you can make a guess of the solution. Then use computations to check if your guess is correct. You can repeat this process until you find the correct solution.

You can use guess and check, along with the following four-step problem solving plan to solve a problem.

Understand • Read and get a general understanding of the problem.

Plan • Make a plan to solve the problem and estimate the solution.

Solve • Use your plan to solve the problem.

Check • Check the reasonableness of your solution.

Example

VETERINARY SCIENCE Dr. Miller saw 40 birds and cats in one day. All together the pets he saw had 110 legs. How many of each type of animal did Dr. Miller see in one day?

Understand You know that Dr. Miller saw 40 birds and cats total. You also know that there were 110 legs in all. You need to find out how many of each type of animal he saw in one day.

Plan Make a guess and check it. Adjust the guess until you get the correct answer.

Solve

Number of birds	Number of cats	Total number of feet
20	20	2(20) + 4(20) = 120
30	10	2(30) + 4(10) = 100
25	15	2(25) + 4(15) = 110

Check 25 birds have 50 feet. 15 cats have 60 feet. Since 50 + 60 is 110, the answer is correct.

Exercise

GEOMETRY In a math class of 26 students, each girl drew a triangle and each boy drew a square. If there were 89 sides in all, how many girls and how many boys were in the class?

Lesson 1-5

1-5 Practice

Problem-Solving Investigation: Guess and Check

Mixed Problem Solving

For Exercises 1 and 2, choose the appropriate method of computation. Then use the method to solve the problem.

1. NUMBERS A number is multiplied by 7. Then 5 is added to the product. The result is 33. What is the number?

2. FOOD Mr. Jones paid $23 for food for his family of seven at the ballpark. Everyone had a drink and either one hot dog or one hamburger. How many hamburgers were ordered?

MENU	
ITEM	**PRICE**
Hot Dog	$2
Hamburger	$3
Drink	$1

Use any strategy to solve Exercises 3–6. Some strategies are shown below.

PROBLEM-SOLVING STRATEGIES
• Guess and Check.
• Find a pattern.

3. PATTERNS What are the next two "words" in the pattern?

ace, bdf, ceg, dfh, egi, _____ , _____

4. GEOMETRY The area of each square is twice the area of the next smaller square drawn in it. If the area of the smallest square is 3 square centimeters, what is the area of the largest square?

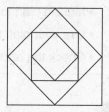

5. ALGEBRA What are the next two numbers in the pattern?

32, 28, 24, 20, _____ , _____

6. MONEY Leeann received $60 for her birthday. The money came in $10 bills and $5 bills. If she received 8 bills, how many of each type did she receive?

7. MONEY Duane has four dimes, half as many nickels as dimes, and three times as many quarters as nickels. How much money does Duane have?

8. LIBRARY Mr. Shuck, the librarian, counted 157 books checked-in during the day. This number was 8 less than 3 times the number of books checked-out that same day. How many books were checked-out that day?

1-6 Study Guide and Intervention

Algebra: Variables and Expressions

To evaluate an algebraic expression you replace each variable with its numerical value, then use the order of operations to simplify.

Example 1 **Evaluate $6x - 7$ if $x = 8$.**

$$6x - 7 = 6(8) - 7 \quad \text{Replace } x \text{ with 8.}$$
$$= 48 - 7 \quad \text{Use the order of operations.}$$
$$= 41 \quad \text{Subtract 7 from 48.}$$

Example 2 **Evaluate $5m - 3n$ if $m = 6$ and $n = 5$.**

$$5m - 3n = 5(6) - 3(5) \quad \text{Replace } m \text{ with 6 and } n \text{ with 5.}$$
$$= 30 - 15 \quad \text{Use the order of operations.}$$
$$= 15 \quad \text{Subtract 15 from 30.}$$

Example 3 **Evaluate $\frac{ab}{3}$ if $a = 7$ and $b = 6$.**

$$\frac{ab}{3} = \frac{(7)(6)}{3} \quad \text{Replace } a \text{ with 7 and } b \text{ with 6.}$$
$$= \frac{42}{3} \quad \text{The fraction bar is like a grouping symbol.}$$
$$= 14 \quad \text{Divide.}$$

Example 4 **Evaluate $x^3 + 4$ if $x = 3$.**

$$x^3 + 4 = 3^3 + 4 \quad \text{Replace } x \text{ with 3.}$$
$$= 27 + 4 \quad \text{Use the order of operations.}$$
$$= 31 \quad \text{Add 27 and 4.}$$

Exercises

Evaluate each expression if $a = 4$, $b = 2$, and $c = 7$.

1. $3ac$

2. $5b^3$

3. abc

4. $5 + 6c$

5. $\frac{ab}{8}$

6. $2a - 3b$

7. $\frac{b^4}{4}$

8. $c - a$

9. $20 - bc$

10. $2bc$

11. $ac - 3b$

12. $6a^2$

13. $7c$

14. $6a - b$

15. $ab - c$

Lesson 1-6

1-6 Practice

Algebra: Variables and Expressions

Evaluate each expression if $r = 5$, $s = 2$, $t = 7$, and $u = 1$.

1. $s + 7$

2. $9 - u$

3. $3t + 1$

4. $5r - 4$

5. $t - s$

6. $u + r$

7. $11t - 7$

8. $6 + 3u$

9. $4r - 10s$

10. $3u^2$

11. $2t^2 - 18$

12. $r^2 + 8$

13. $\dfrac{s}{2}$

14. $\dfrac{30}{r}$

15. $\dfrac{(3 + u)^2}{8}$

Evaluate each expression if $a = 4.1$, $b = 5.7$, and $c = 0.3$.

16. $a + b - c$

17. $10 - (a + b)$

18. $b - c + 2$

19. **MOON** The expression $\dfrac{w}{6}$ gives the weight of an object on the Moon in pounds with a weight of w pounds on Earth. What is the weight of a space suit on the Moon if the space suit weighs 178.2 pounds on Earth?

20. **Complete the table.**

Pounds (p)	Ounces ($16p$)
1	16
2	32
3	
4	
5	

1-7 Study Guide and Intervention

Algebra: Equations

- An **equation** is a sentence in mathematics that contains an equals sign, =.
- The **solution** of an equation is the value that when substituted for the variable makes the equation true.

Example 1 Solve $23 + y = 29$ mentally.

$23 + y = 29$	Write the equation.
$23 + 6 = 29$	You know that $23 + 6$ is 29.
$29 = 29$	Simplify.

The solution is 6.

Example 2

TRAVEL On their annual family vacation, the Wilsons travel 790 miles in two days. If on the first day they travel 490 miles, how many miles must they drive on the second day to reach their destination?

The total distance to travel in two days is 790 miles.

Let m represent the distance to travel on day two.

$m + 490 = 790$

$m + 490 = 790$	Write the equation.
$300 + 490 = 790$	Replace m with 300 to make the equation true.
$790 = 790$	Simplify.

The number 300 is the solution. The distance the Wilsons must travel on day two is 300 miles.

Exercises

Solve each equation mentally.

1. $k + 7 = 15$ 2. $g - 8 = 20$ 3. $6y = 24$

4. $\dfrac{a}{3} = 9$ 5. $\dfrac{x}{6} = 9$ 6. $8 + r = 24$

7. $12 \cdot 8 = h$ 8. $n \div 11 = 8$ 9. $48 \div 12 = x$

10. $h - 12 = 24$ 11. $19 + y = 28$ 12. $9f = 90$

Define a variable. Then write and solve an equation.

13. **MONEY** Aaron wants to buy a video game. The game costs $15.50. He has $10.00 saved from his weekly allowance. How much money does he need to borrow from his mother in order to buy the video game?

1-7 Practice

Algebra: Equations

Solve each equation mentally.

1. $a + 5 = 14$

2. $7 + y = 24$

3. $t - 13 = 33$

4. $b - 17 = 11$

5. $12 - r = 0$

6. $x + 18 = 59$

7. $63 = 9g$

8. $8d = 96$

9. $n = \dfrac{42}{7}$

10. $9 = \dfrac{z}{7}$

11. $10 = h \div 4$

12. $55 \div m = 11$

13. $1.2 + k = 3.0$

14. $2.7 = f - 1.1$

15. $v - 0.5 = 0.2$

16. $12.6 - c = 7.0$

17. $8.8 + j = 18.7$

18. $w + 13.5 = 16.0$

19. WEATHER The temperature was 78°F. A cold front moved in, and the temperature dropped to 54°F. Solve the equation $78 - d = 54$ to find the drop in temperature.

20. HOBBIES Elissa can cut out the pieces of cloth to make four pillows in one hour. Solve the equation $4h = 20$ to find how many hours Elissa needs to cut cloth for 20 pillows.

21. BOWLING Jean Conrad is an amateur bowler with an average score of 187. She recently bowled a perfect 300 score. Write an equation that can be used to find how much the perfect score was above her average score and then solve the equation.

1-8 Study Guide and Intervention

Algebra: Properties

Lesson 1-8

Property	Arithmetic	Algebra
Distributive Property	$5(3 + 4) = 5(3) + 5(4)$	$a(b + c) = a(b) + a(c)$
Commutative Property of Addition	$5 + 3 = 3 + 5$	$a + b = b + a$
Commutative Property of Multiplication	$5 \times 3 = 3 \times 5$	$a \times b = b \times a$
Associative Property of Addition	$(2 + 3) + 4 = 2 + (3 + 4)$	$(a + b) + c = a + (b + c)$
Associative Property of Multiplication	$(4 \times 5) \times 6 = 4 \times (5 \times 6)$	$(a \times b) \times c = a \times (b \times c)$
Identity Property of Addition	$5 + 0 = 5$	$a + 0 = a$
Identity Property of Multiplication	$5 \times 1 = 5$	$a \times 1 = a$

Example 1 Use the Distributive Property to write $6(4 + 3)$ as an equivalent expression. Then evaluate the expression.

$$6(4 + 3) = 6 \cdot 4 + 6 \cdot 3 \qquad \text{Apply the Distributive Property.}$$
$$= 24 + 18 \qquad \text{Multiply.}$$
$$= 42 \qquad \text{Add.}$$

Example 2 Name the property shown by each statement.

$5 \times 4 = 4 \times 5$ Commutative Property of Multiplication

$12 + 0 = 12$ Identity Property of Addition

$7 + (6 + 3) = (7 + 6) + 3$ Associative Property of Addition

Exercises

Use the Distributive Property to write each expression as an equivalent expression. Then evaluate the expression.

1. $5(7 + 2)$ **2.** $4(9 + 1)$ **3.** $2(6 + 7)$

Name the property shown by each statement.

4. $9 \times 1 = 9$ **5.** $7 \times 3 = 3 \times 7$

6. $(7 + 8) + 2 = 7 + (8 + 2)$ **7.** $6(3 + 2) = 6(3) + 6(2)$

8. $15 + 12 = 12 + 15$ **9.** $1 \times 20 = 20$

10. $(9 \times 5) \times 2 = 9 \times (5 \times 2)$ **11.** $3 = 0 + 3$

1-8 Practice

Algebra: Properties

Use the Distributive Property to evaluate each expression.

1. $4(5 + 7)$ **2.** $6(3 + 1)$ **3.** $(10 + 8)2$

4. $5(8 - 3)$ **5.** $7(4 - 1)$ **6.** $(9 - 2)3$

Name the property shown by each statement.

7. $7 + (6 + t) = (7 + 6) + t$ **8.** $23 \cdot 15 = 15 \cdot 23$ **9.** $0 + x = x$

10. $3(g + 7) = 3 \cdot g + 3 \cdot 7$ **11.** $8 \times 1 = 8$ **12.** $y + 11 = 11 + y$

13. $5(w + 1) = (w + 1)5$ **14.** $(4 \cdot d) \cdot 1 = 4 \cdot (d \cdot 1)$ **15.** $(6 + 2)7 = (6)7 + (2)7$

Use one or more properties to rewrite each expression as an equivalent expression that does not use parentheses.

16. $(b + 3) + 6$ **17.** $7(5x)$ **18.** $4(a + 4)$

19. $7 + (3 + t)$ **20.** $(2z)0$ **21.** $(9 + k)5$

22. $8(y - 5) + y$ **23.** $(h + 2)3 - 2h$

24. GROCERY A grocery store sells an imported specialty cheesecake for \$11 and its own store-baked cheesecake for \$5. Use the Distributive Property to mentally find the total cost for 6 of each type of cheesecake.

25. CHECKING ACCOUNT Mr. Kenrick balances his checking account statement each month two different ways as shown by the equation, $(b + d) - c = b + (d - c)$, where b is the previous balance, d is the amount of deposits made, and c is the amount of checks written. Name the property that Mr. Kenrick uses to double check his arithmetic.

26. SPEED A train is traveling at a speed of 65 miles per hour. The train travels for one hour. What property is used to solve this problem as shown by the statement $65 \cdot 1 = 65$?

1-9 Study Guide and Intervention

Algebra: Arithmetic Sequences

An **arithmetic sequence** is a list in which each term is found by adding the same number to the previous term. 1, 3, 5, 7, 9, ...
+2 +2 +2 +2

Example 1 Describe the relationship between terms in the arithmetic sequence 17, 23, 29, 35, ... Then write the next three terms in the sequence.

17, 23, 29, 35, ...   Each term is found by adding 6 to the previous term.
+6 +6 +6 $35 + 6 = 41$ $41 + 6 = 47$ $47 + 6 = 53$

The next three terms are 41, 47, and 53.

Example 2

MONEY Brian's parents have decided to start giving him a monthly allowance for one year. Each month they will increase his allowance by $10. Suppose this pattern continues. What algebraic expression can be used to find Brian's allowance after any given number of months? How much money will Brian receive for allowance for the 10th month?

Make a table to display the sequence.

Position	Operation	Value of Term
1	$1 \cdot 10$	10
2	$2 \cdot 10$	20
3	$3 \cdot 10$	30
n	$n \cdot 10$	$10n$

Each term is 20 times its position number. So, the expression is $10n$.
How much money will Brian earn after 10 months?
$10n$ Write the expression.
$10(10) = 100$ Replace n with 10

So, for the 10th month Brian will receive $100.

Exercises

Describe the relationship between terms in the arithmetic sequences. Write the next three terms in the sequence.

1. 2, 4, 6, 8, ...   2. 4, 7, 10, 13, ...   3. 0.3, 0.6, 0.9, 1.2, ...

4. 200, 212, 224, 236, ...   5. 1.5, 2.0, 2.5, 3.0, ...   6. 12, 19, 26, 33, ...

7. **SALES** Mama's bakery just opened and is currently selling only two types of pastry. Each month, Mama's bakery will add two more types of pastry to their menu. Suppose this pattern continues. What algebraic expression can be used to find the number of pastries offered after any given number of months? How many pastries will be offered in one year?

Lesson 1-9

1-9 Practice

Algebra: Arithmetic Sequences

Describe the relationship between the terms in each arithmetic sequence. Then write the next three terms in each sequence.

1. 0, 5, 10, 15, ...

2. 1, 3, 5, 7, ...

3. 18, 27, 36, 45, ...

4. 7, 19, 31, 43, ...

5. 8, 18, 28, 38, ...

6. 25, 26, 27, 28, ...

7. 0.4, 0.8, 1.2, 1.6, ...

8. 3.7, 3.7, 3.7, 3.7, ...

9. 5.1, 6.2, 7.3, 8.4, ...

10. 17, 31, 45, 59, ...

11. 30, 50, 70, 90, ...

12. 14, 41, 68, 95, ...

In a *geometric sequence*, each term is found by multiplying the previous term by the same number. Write the next three terms of each geometric sequence.

13. 5, 10, 20, 40, ...

14. 3, 9, 27, 81, ...

15. 2, 8, 32, 128, ...

NUMBER SENSE   **Find the 40th term in each arithmetic sequence.**

16. 4, 8, 12, 16, ...

17. 13, 26, 39, 52, ...

18. 6, 12, 18, 24, ...

19. GEOMETRY The lengths of the sides of a 6-sided polygon are in arithmetic sequence. The length of the shortest side is 3 meters. If the length of the next longer side is 5 meters, what is the length of the longest side?

20. FREE FALLING OBJECT A free falling object increases speed by a little over 22 miles per hour each second. The arithmetic sequence 22, 44, 66, ..., represents the speed after each second, in miles per hour, of a dropped object. How fast is a rock falling after 8 seconds if it is dropped over the side of a cliff?

1-10 Study Guide and Intervention

Algebra: Equations and Functions

The solution of an equation with two variables consists of two numbers, one for each variable that makes the equation true. When a relationship assigns exactly one output value for each input value, it is called a function. Function tables help to organize input numbers, output numbers, and function rules.

Example 1 Complete a function table for $y = 5x$. Then state the domain and range.

Choose four values for x. Substitute the values for x into the expression. Then evaluate to find the y value.

x	$5x$	y
0	5(0)	0
1	5(1)	5
2	5(2)	10
3	5(3)	15

The domain is {0, 1, 2, 3}. The range is {0, 5, 10, 15}.

Exercises

Complete the following function tables. Then state the domain and range.

1. $y = x + 4$

x	$x + 4$	y
0	0+4	4
1	1+4	5
2	2+4	6
3	3+4	7

2. $y = 10x$

x	$10x$	y
1	10(1)	10
2	10(2)	20
3	10(3)	30
4	10(4)	40

3. $y = x - 1$

x	$x - 1$	y
2	2-1	1
3	3-1	2
4	4-1	3
5	5-1	4

4. $y = 3x$

x	$3x$	y
10	3(10)	30
11	3(11)	33
12	3(12)	36
13	3(13)	39

Lesson 1-10

1-10 Practice

Algebra: Equations and Functions

Complete each function table. Then identify the domain and range.

1. $y = 5x$

x	$5x$	y
1		
2		
3		
4		

2. $y = 8x$

x	$8x$	y
1		
2		
3		
4		

3. $y = 7x$

x	$7x$	y
3		
4		
5		
6		

4. $y = x - 2$

x	$x - 2$	y
2		
3		
4		
5		

5. $y = x + 3$

x	$x + 3$	y
2		
3		
4		
5		

6. $y = x + 0.75$

x	$x + 0.75$	y
0		
1		
2		
3		

7. PRODUCTION A car manufacturer makes 15,000 hybrid cars a month. Using the function table, find the number of hybrid cars produced after 3, 6, 9, and 12 months.

m	$15,000m$	P
3		
6		
9		
12		

8. SUNSPOTS The changing activity of sunspots, which are cooler and darker areas of the sun, occur in 11-year cycles. Use the function $y = 11c$ to find the numbers of years necessary to complete 1, 2, 3, and 4 sunspot cycles.

2-1 Study Guide and Intervention

Integers and Absolute Value

Integers less than zero are **negative integers**. Integers greater than zero are **positive integers**.

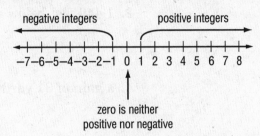

The **absolute value** of an integer is the distance the number is from zero on a number line. Two vertical bars are used to represent absolute value. The symbol for absolute value of 3 is $|3|$.

Example 1 **Write an integer that represents 160 feet below sea level.**

Because it represents *below* sea level, the integer is -160.

Example 2 **Evaluate $|-2|$.**

On the number line, the graph of -2 is
2 units away from 0. So, $|-2| = 2$.

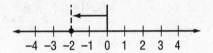

Exercises

Write an integer for each situation.

1. 12°C above 0

2. a loss of $24

3. a gain of 20 pounds

4. falling 6 feet

Evaluate each expression.

5. $|12|$

6. $|-150|$

7. $|-8|$

8. $|75|$

9. $|-19|$

10. $|84|$

Lesson 2-1

2-1 Practice

Integers and Absolute Value

Write an integer for each situation.

1. a profit of $12

2. 1,440 feet below sea level

3. 22°F below 0

4. a gain of 31 yards

Graph each set of integers on a number line.

5. $\{-5, 0, 5\}$

6. $\{-3, -2, 1, -4\}$

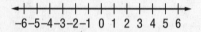

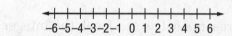

Evaluate each expression.

7. $|-11|$

8. $|-5| + 8$

9. $|-4| - |-4|$

10. $|12| \div 2 \times |-5|$

11. $|-4| + 7 - |3|$

12. $9 + |-6| \div 1^2$

13. HEALTH A veterinarian recommends that a St. Bernard lose weight. Write an integer to describe the dog losing 25 pounds.

14. GEOGRAPHY Mount Kilimanjaro is the highest peak in Africa. Write an integer to represent the elevation of Mount Kilimanjaro of 5,895 meters above sea level.

15. ECONOMY Gasoline prices occasionally fluctuate during a two month period of time. Prices increased 34 cents per gallon during the month of April and decreased 17 cents per gallon during the month of May. What integers can be used to describe each change in price?

2-2 Study Guide and Intervention

Comparing and Ordering Integers

When two numbers are graphed on a number line, the number to the left is always less than (<) the number to the right. The number to the right is always greater than (>) the number to the left.

Model

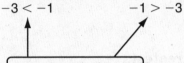

Words   −3 is less than −1.   −1 is greater than −1.

Symbols   −3 < −1   −1 > −3

The symbol points to the lesser number.

Example 1   **Replace the ● with < or > to make −1 ● −6 a true sentence.**

Graph each integer on a number line.

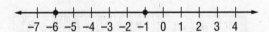

Since −1 is to the right of −6, −1 > −6.

Example 2   **Order the integers 2, −3, 0, −5 from least to greatest.**

To order the integers, graph them on a number line.

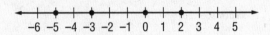

Order the integers by reading from left to right: −5, −3, 0, 2.

Exercises

1. Replace the ● with < or > to make −5 ● −10 a true sentence.

2. Order −1, 5, −3, and 2 from least to greatest.

3. Order 0, −4, −2, and 7 from greatest to least.

4. Order −3, |−2|, 4, 0, and −5 from greatest to least.

Lesson 2-2

2-2 Practice

Comparing and Ordering Integers

Replace each ● with < or > to make a true sentence.

1. −5 ● 1

2. −27 ● −31

3. 7 ● 0

4. 4 ● −11

5. 7 ● −7

6. 12 ● −14

7. −54 ● −31

8. −49 ● 3

9. −1 ● 2

Order the integers in each set from least to greatest.

10. {−4, 4, −1, 7, 2}

11. {8, −5, 0, 1, −2}

12. {11, −17, 12, −9, 3, −1}

Replace each ● with <, >, or = to make a true sentence.

13. 4 ● |−4|

14. |−27| ● |−31|

15. 12 ● |−18|

16. **ANALYZE TABLES** Elements melt at different temperatures. Five elements and their melting points in °C, are listed in the table. Order the elements from the lowest melting point to the highest melting point.

Element	Melting Point °C
Carbon	3,500
Helium	−272
Mercury	−39
Oxygen	−218
Sodium	98

Source: science.co.il

BUSINESS For Exercises 17 and 18, use the information in the table. It shows the net profit or loss of a used-car dealership during the spring and summer months of a recent year.

Month	March	April	May	June	July	August	September
Net Profit Or Loss	$8,500	$1,800	−$2,300	$300	−$1,000	$9,400	$2,500

17. Order the months from the lowest net value to the highest.

18. Which net value is the middle, or **median**, value?

2-3 Study Guide and Intervention

The Coordinate Plane

Lesson 2-3

The **coordinate plane** is used to locate points. The horizontal number line is the **x-axis**. The vertical number line is the **y-axis**. Their intersection is the **origin**.

Points are located using **ordered pairs**. The first number in an ordered pair is the **x-coordinate**; the second number is the **y-coordinate**.

The coordinate plane is separated into four sections called **quadrants**.

Example 1 Name the ordered pair for point P. Then identify the quadrant in which P lies.

- Start at the origin.
- Move 4 units left along the x-axis.
- Move 3 units up on the y-axis.
  The ordered pair for point P is (−4, 3).
  P is in the upper left quadrant or quadrant II.

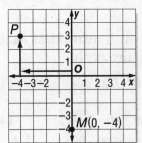

Example 2 Graph and label the point M(0, −4).

- Start at the origin.
- Move 0 units along the x-axis.
- Move 4 units down on the y-axis.
- Draw a dot and label it M(0, −4).

Exercises

Name the ordered pair for each point graphed at the right. Then identify the quadrant in which each point lies.

1. P   2. Q

3. R   4. S

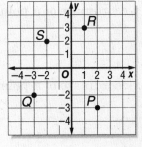

Graph and label each point on the coordinate plane.

5. A(−1, 1)   6. B(0, −3)

7. C(3, 2)   8. D(−3, −1)

9. E(1, −2)   10. F(1, 3)

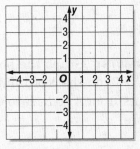

2-3 Practice

The Coordinate Plane

Write the ordered pair for each point graphed at the right. Then name the quadrant or axis on which each point is located.

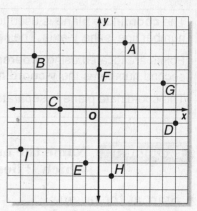

1. A
2. B
3. C

4. D
5. E
6. F

7. G
8. H
9. I

Graph and label each point on the coordinate plane at the right.

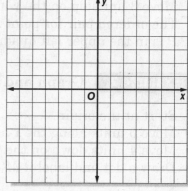

10. $J(2, 2)$
11. $K(-3, 4)$
12. $L(-4, 1)$

13. $M(-3, -3)$
14. $N(1, -4)$
15. $O(0, 0)$

16. $P(4, 5)$
17. $Q(4, -3)$
18. $R(-6, -5)$

Determine whether each statement is *sometimes*, *always*, or *never* true. Explain or give a counterexample to support your answer.

19. The y-coordinate of a point in quadrant II is negative.

20. The x-coordinate of a point on the y-axis is zero.

21. In quadrants I and III, the x-coordinate of a point is positive.

22. **GEOMETRY** Graph the points $A(-3, -1)$, $B(0, 4)$, $C(4, 3)$, and $D(1, -2)$ on the coordinate plane at the right. Connect the points from A to B, B to C, C to D, and D to A. Name the figure.

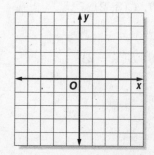

3-1 Study Guide and Intervention

Writing Expressions and Equations

The table below shows phrases written as mathematical expressions.

Phrases	Expression	Phrases	Expression
9 more than a number the sum of 9 and a number a number plus 9 a number increased by 9 the total of x and 9	$x + 9$	4 subtracted from a number a number minus 4 4 less than a number a number decreased by 4 the difference of h and 4	$h - 4$
Phrases	**Expression**	**Phrases**	**Expression**
6 multiplied by g 6 times a number the product of g and 6	$6g$	a number divided by 5 the quotient of t and 5 divide a number by 5	$\dfrac{t}{5}$

The table below shows sentences written as an equation.

Sentences	Equation
Sixty less than three times the amount is $59. Three times the amount less 60 is equal to 59. 59 is equal to 60 subtracted from three times a number. A number times three minus 60 equals 59.	$3n - 60 = 59$

Exercises

Write each phrase as an algebraic expression.

1. 7 less than m

2. the quotient of 3 and y

3. the total of 5 and c

4. the difference of 6 and r

5. n divided by 2

6. the product of k and 9

Write each sentence as an algebraic equation.

7. A number increased by 7 is 11.

8. The price decreased by $4 is $29.

9. Twice as many points as Bob would be 18 points.

10. After dividing the money 5 ways, each person got $67.

11. Three more than 8 times as many trees is 75 trees.

12. Seven less than a number is 15.

37

3-1 Practice

Writing Expressions and Equations

Write each phrase as an algebraic expression.

1. the product of -5 and x

2. twenty increased by k

3. five inches more than the height

4. one fourth of y

5. Bill's weight decreased by eighteen

6. the quotient of 3 and a number

7. five less than four times as many women

8. $60 more than the rent payment

9. 9 minutes less than Chang's time

10. three more pancakes than his brother ate

Write each sentence as an algebraic equation.

11. Five times the number of books is 95.

12. The difference of nine and a number is nine.

13. The sum of the average and four is -6.

14. Three meters longer than the pool is 8.

15. Twelve less a number is 40.

16. The product of seven and Lynn's age is 28.

For Exercises 17 and 18, write an equation that models each situation.

17. **FURNITURE** The width of a bookshelf is 2 feet shorter than the height. If the width is 1.5 feet, what is the height of the bookshelf?

18. **SPORTS** The circumference of a basketball, the distance around, is about three times the circumference of a softball. If the circumference of the basketball is 75 centimeters, what is the circumference of a softball?

GEOMETRY For Exercises 19 and 20, describe the relationship that exists between the base and the height of each triangle.

19. The base is b, and the height is $b - 4$.

20. The height is h, and the base is $2h$.

3-2 Study Guide and Intervention

Solving Addition and Subtraction Equations

Remember, equations must always remain balanced. If you subtract the same number from each side of an equation, the two sides remain equal. Also, if you add the same number to each side of an equation, the two sides remain equal.

Example 1 Solve $x + 5 = 11$. Check your solution.

$x + 5 = 11$	Write the equation.
$\underline{-5 = -5}$	Subtract 5 from each side.
$x = 6$	Simplify.

Check
$x + 5 = 11$	Write the equation.
$6 + 5 \overset{?}{=} 11$	Replace x with 6.
$11 = 11$ ✓	This sentence is true.

The solution is 6.

Example 2 Solve $15 = t - 12$. Check your solution.

$15 = t - 12$	Write the equation.
$\underline{+12 = +12}$	Add 12 to each side.
$27 = t$	Simplify.

Check
$15 = t - 12$	Write the equation.
$15 \overset{?}{=} 27 - 12$	Replace t with 27.
$15 = 15$ ✓	This sentence is true.

The solution is 27.

Exercises

Solve each equation. Check your solution.

1. $h + 3 = 14$
$h + 3 - 3 = 14 - 3$
$h = 11$

2. $m + 8 = 22$
$m + 8 - 8 = 22 - 8$
$m = 14$

3. $p + 5 = 15$
$p + 5 - 5 = 15 - 5$
$p = 10$

4. $17 = y + 8$
$17 + 8 = y + 8 - 8$
$9 = 4$

5. $w + 4 = -1$
$w + 4 - 4 = -1 + 4$
$w = 3$

6. $k + 5 = -3$
$k + 5 - 5 = -3 - 5$
$k = -8$

7. $25 = 14 + r$
$25 - 14 = 14 - 14 + r$
$11 = r$

8. $57 + z = 97$
$57 - 57 + z = 97 - 57$
$z = 40$

9. $b - 3 = 6$
$b - 3 + 3 = 6 + 3$
$b = 9$

10. $7 = c - 5$
$7 + 5 = c - 5 + 5$
$12 = c$

11. $j - 12 = 18$
$j - 12 + 12 = 18 + 12$
$j = 32$

12. $v - 4 = 18$
$v - 4 + 4 = 18 + 4$
$v = 22$

13. $-9 = w - 12$
$-9 + 12 = w - 12 + 12$
$3 = w$

14. $y - 8 = -12$
$y - 8 + 8 = -12 + 8$
$y = -4$

15. $14 = f - 2$
$14 + 2 = f - 2 + 2$
$16 = f$

16. $23 = n - 12$
$23 + 12 = n - 12 + 12$
$35 = n$

Lesson 3-2

3-2 Practice

Solving Addition and Subtraction Equations

Solve each equation. Check your solution.

1. $a + 4 = 11$

2. $6 = g + 8$

3. $x - 3 = -2$

4. $k + 8 = 3$

5. $j + 0 = 9$

6. $12 + y = 15$

7. $h - 4 = 0$

8. $m - 7 = 1$

9. $w + 5 = 4$

10. $b - 28 = 33$

11. $45 + f = 48$

12. $n + 7.1 = 8.6$

13. $-14 + t = 26$

14. $d - 3.03 = 2$

15. $10 = z + 15$

16. $c - 5.3 = -6.4$

17. $35 + p = 77$

18. $-15 = -15 + u$

For Exercises 19 and 20, write an equation. Then solve the equation.

19. CAFFEINE   A cup of brewed tea has 54 milligrams less caffeine than a cup of brewed coffee. If a cup of tea has 66 milligrams of caffeine, how much caffeine is in a cup of coffee?

20. GEOMETRY   The sum of the measures of the angles of a trapezoid is 360°. Find the missing measure.

3-3 Study Guide and Intervention

Solving Multiplication Equations

If each side of an equation is divided by the same non-zero number, the resulting equation is equivalent to the given one. You can use this property to solve equations involving multiplication and division.

Example 1 Solve $45 = 5x$. Check your solution.

$45 = 5x$ Write the equation.

$\dfrac{45}{5} = \dfrac{5x}{5}$ Divide each side of the equation by 5.

$9 = x$ $45 \div 5 = 9$

Check $45 = 5x$ Write the original equation.

$45 \overset{?}{=} 5(9)$ Replace x with 9. Is this sentence true?

$45 = 45$ ✓

The solution is 9.

Example 2 Solve $-21 = -3y$. Check your solution.

$-21 = -3y$ Write the equation.

$\dfrac{-21}{-3} = \dfrac{-3y}{-3}$ Divide each side by -3.

$7 = y$ $-21 \div (-3) = 7$

Check $-21 = -3y$ Write the original equation.

$-21 \overset{?}{=} -3(7)$ Replace y with 7. Is this sentence true?

$-21 = -21$ ✓

The solution is 7.

Exercises

Solve each equation. Then check your solution.

1. $8q = 56$ **2.** $4p = 32$ **3.** $42 = 6m$ **4.** $104 = 13h$

5. $-6n = 30$ **6.** $-18x = 36$ **7.** $48 = -8y$ **8.** $72 = -3b$

9. $-9a = -45$ **10.** $-12m = -120$ **11.** $-66 = -11t$ **12.** $-144 = -9r$

13. $3a = 4.5$ **14.** $2h = 3.8$ **15.** $4.9 = 0.7k$ **16.** $9.75 = 2.5z$

Lesson 3-3

3-3 Practice

Solving Multiplication Equations

Solve each equation. Check your solution.

1. $8e = 32$

2. $4v = -8$

3. $7k = -7$

4. $18 = 3y$

5. $4j = 0$

6. $-11x = -44$

7. $5a = 5$

8. $-1c = 8$

9. $15 = 5b$

10. $-2w = -14$

11. $9f = 45$

12. $13m = -26$

13. $1.4t = 2.8$

14. $0.9g = 5.4$

15. $2.5 = 0.5h$

16. $3.74 = 1.7d$

17. $4.1z = 16.81$

18. $5.2q = 3.64$

For Exercises 19 and 20, write an equation. Then solve the equation.

19. **TRAVEL**   A cheetah can travel at an amazing speed of 32 meters per second when chasing its prey. At that rate, how long would it take the cheetah to run 2,000 meters?

20. **AUTO LOAN**   Mrs. Kim borrowed $1,350 to buy a used automobile. If she repays $75 a month, how many months will it take to pay back the loan?

3-4 Study Guide and Intervention

Problem-Solving Investigation: Work Backward

By working backward from where you end to where you began, you can solve problems. Use the four-step problem solving model to stay organized when working backward.

Example 1 Jonah put half of his birthday money into his savings account. Then he paid back the $10 that he owed his brother for dance tickets. Lastly, he spent $3 on lunch at school. At the end of the day he was left with $12. How much money did Jonah receive for his birthday?

Understand	You know that he had $12 left and the amounts he spent throughout the day. You need to find out how much money he received for his birthday.
Plan	Start with the amount of money he was left with and work backward.
Solve	He had $12 left. Undo the $3 he spent on lunch. Undo the $10 he gave back to his brother. Undo the half put into his savings account. So, Jonah received $50 for his birthday.
Check	Assume that Jonah receive $50 for his birthday. After putting half into his savings account he had $50 ÷ 2 or $25. Then he gave $10 to his brother for dance tickets, so he had $25 − $10 or $15. Lastly, he spent $3 on lunch at school, so he had $15 − $3, or $12. So, our answer of $50 is correct.

In the Solve row the computation shown:

$$\begin{array}{r} 12 \\ +\ 3 \\ \hline 15 \\ +\ 10 \\ \hline 25 \\ \times\ 2 \\ \hline 50 \end{array}$$

Exercises

Solve each problem by using the work backward strategy.

1. On Monday everyone was present in Mr. Miller's class. At 12:00, 5 students left early for doctors' appointments. At 1:15, half of the remaining students went to an assembly. Finally, at 2:00, 6 more students left for a student council meeting. At the end of the day, there were only 5 students in the room. Assuming that no students returned after having left, how many students are in Mr. Miller's class?

2. Jordan was trading baseball cards with some friends. He gave 15 cards to Tommy and got 3 back. He gave two-thirds of his remaining cards to Elaine and kept the rest for himself. When he got home he counted that he had 25 cards. How many baseball cards did Jordan start with?

Lesson 3-4

3-4 Practice

Problem-Solving Investigation: Work Backward

Mixed Problem Solving

Use the work backward strategy to solve Exercises 1 and 2.

1. **NUMBER THEORY** A number is divided by 5. Then 3 is added to the quotient. After subtracting 10, the result is 30. What is the number?

2. **COUPONS** Kendra used 35 cents more in coupons at the store than Leanne. Leanne used 75 cents less than Teresa, who used 50 cents more than Jaclyn. Jaclyn used 40 cents in coupons. What was the value of the coupons Kendra used?

Use any strategy to solve Exercises 3–6. Some strategies are shown below.

PROBLEM-SOLVING STRATEGIES
• Look for a pattern
• Guess and check.
• Work backward

3. **PATTERNS** What are the next three numbers in the following pattern?

2, 3, 5, 9, 17, 33, . . .

4. **AGES** Mr. Gilliam is 3 years younger than his wife. The sum of their ages is 95. How old is Mr. Gilliam?

5. **GRAND CANYON** The elevation of the North Rim of the Grand Canyon is 2,438 meters above sea level. The South Rim averages 304 meters lower than the North Rim. What is the average elevation of the South Rim?

6. **WATER BILL** The water company charges a residential customer $41 for the first 3,000 gallons of water used and $1 for every 200 gallons used over 3,000 gallons. If the water bill was $58, how many gallons of water were used?

3-5 Study Guide and Intervention

Solving Two-Step Equations

To solve two-step equations, you need to add or subtract first. Then divide to solve the equation.

Example 1 Solve $7v - 3 = 25$. Check your solution.

$$
\begin{array}{ll}
7v - 3 = 25 & \text{Write the equation.} \\
\underline{+3 = +3} & \text{Add 3 to each side.} \\
7v = 28 & \text{Simplify.} \\
\dfrac{7v}{7} = \dfrac{28}{7} & \text{Divide each side by 7.} \\
v = 4 & \text{Simplify.}
\end{array}
$$

$$
\begin{array}{lll}
\textbf{Check} & 7v - 3 = 25 & \text{Write the original equation.} \\
& 7(4) - 3 \stackrel{?}{=} 25 & \text{Replace } v \text{ with 4.} \\
& 28 - 3 \stackrel{?}{=} 25 & \text{Multiply.} \\
& 25 = 25\ \checkmark & \text{The solution checks.}
\end{array}
$$

The solution is 4.

Example 2 Solve $-10 = 8 + 3x$. Check your solution.

$$
\begin{array}{ll}
-10 = 8 + 3x & \text{Write the equation.} \\
\underline{-8 = -8} & \text{Subtract 8 from each side.} \\
-18 = 3x & \text{Simplify.} \\
\dfrac{-18}{3} = \dfrac{3x}{3} & \text{Divide each side by 3.} \\
-6 = x & \text{Simplify.}
\end{array}
$$

$$
\begin{array}{lll}
\textbf{Check} & -10 = 8 + 3x & \text{Write the original equation.} \\
& -10 \stackrel{?}{=} 8 + 3(-6) & \text{Replace } x \text{ with } -6. \\
& -10 \stackrel{?}{=} 8 + (-18) & \text{Multiply.} \\
& -10 = -10\ \checkmark & \text{The solution checks.}
\end{array}
$$

The solution is -6.

Exercises

Solve each equation. Check your solution.

1. $4y + 1 = 13$ 2. $6x + 2 = 26$ 3. $-3 = 5k + 7$ 4. $6n + 4 = -26$

5. $7 = -3c - 2$ 6. $-8p + 3 = -29$ 7. $-5 = -5t - 5$ 8. $-9r + 12 = -24$

9. $11 + 7n = 4$ 10. $35 = 7 + 4b$ 11. $15 + 2p = 9$ 12. $49 = 16 + 3y$

13. $2 = 4t - 14$ 14. $-9x - 10 = 62$ 15. $30 = 12z - 18$ 16. $7 + 4g = 7$

17. $24 + 9x = -3$ 18. $50 = 16q + 2$ 19. $3c - 2.5 = 4.1$ 20. $9y + 4.8 = 17.4$

Lesson 3-5

3-5 Practice

Solving Two-Step Equations

Solve each equation. Check your solution.

1. $4h + 6 = 30$

2. $7y + 5 = -9$

3. $-3t + 6 = 0$

4. $-8 + 8g = 56$

5. $5k - 7 = -7$

6. $19 + 13x = 32$

7. $-5b - 12 = -2$

8. $-1n + 1 = 11$

9. $9f + 15 = 51$

10. $5d - 3.3 = 7.2$

11. $3 = 0.2m - 7$

12. $1.3z + 1.5 = 5.4$

13. KITTENS Kittens weigh about 100 grams when born and gain 7 to 15 grams per day. If a kitten weighed 100 grams at birth and gained 8 grams per day, in how many days will the kitten triple its weight?

14. TEMPERATURE Room temperature ranges from 20°C to 25°C. Find the range of room temperature in °F. Use the formula, $F - 32 = 1.8C$, to convert from the Celsius scale to the Fahrenheit scale.

3-6 Study Guide and Intervention

Measurement: Perimeter and Area

The distance around a geometric figure is called the **perimeter**.

To find the perimeter of any geometric figure, you can use addition or a formula.

The perimeter of a rectangle is twice the length ℓ plus twice the width w.

$$P = 2\ell + 2w$$

Example 1   Find the perimeter of the figure at right.

$P = 105 + 105 + 35 + 35$ or 280

The perimeter is 280 inches.

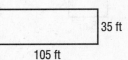
35 ft
105 ft

The measure of the surface enclosed by a geometric figure is called the **area**.

The area of a rectangle is the product of the length ℓ and width w.

$$A = \ell \cdot w$$

Example 2   Find the area of the rectangle.

$A = \ell \cdot w$
$\quad = 24 \cdot 12$ or 288

The area is 288 square centimeters.

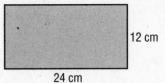
12 cm
24 cm

Exercises

Find the perimeter of each figure.

1.
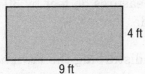
7 cm
33 cm

2.
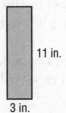
42 m
14 m

Find the perimeter and area of each rectangle.

3.
4 ft
9 ft

4.
11 in.
3 in.

5. $\ell = 8$ ft, $w = 5$ ft

6. $\ell = 3.5$ m, $w = 2$ m

7. $\ell = 8$ yd, $w = 4\frac{1}{3}$ yd

8. $\ell = 29$ cm, $w = 7.3$ cm

Lesson 3-6

3-6 Practice

Measurement: Perimeter and Area

Find the perimeter of each rectangle.

1.
15 m

5 m

2.
2.9 mi

2.8 mi

3.
1 yd

0.5 yd

Find the area of each rectangle.

4.
26 in.

11 in.

5.
8.5 ft

7.6 ft

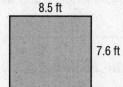

6.
12 cm

10 cm

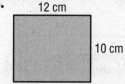

Find the missing side.

7. $P = 83.4$ km, $\ell = 27.8$ km

8. $A = 337.68$ yd^2, $w = 60.3$ yd

LAWN CARE   For Exercises 9 and 10, use the following information.
Yuri's dad needs to fertilize the grass in the yard. The back yard measures 55 feet by 30 feet, while the front yard is a square with a length of 42 feet on each side.

9. Yuri's dad wants to rope off the two areas to keep people from disturbing the lawn after he fertilizes the grass. How much rope will he need to go around both areas?

10. If a bag of fertilizer covers 600 square feet of lawn, how many bags of fertilizer will Yuri's dad need to fertilize the front and back yards?

4-1 Study Guide and Intervention

Prime Factorization

Lesson 4-1

A whole number is **prime** if it has exactly two factors, 1 and itself. A whole number is **composite** if it is greater than one and has more than two factors. To determine the **prime factorization** of a number, use a **factor tree**.

Example 1 Determine whether each number is *prime* or *composite*.

   a. 11   b. 24

   a. The number 11 has only two factors, 1 and 11, so it is prime.

   b. The number 24 has 8 factors, 1, 2, 3, 4, 6, 8, 12, and 24. So, it is composite.

Example 2 Determine the prime factorization of 48.

Use a factor tree.

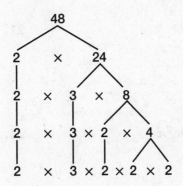

The prime factorization of 48 is $2 \times 2 \times 2 \times 2 \times 3$ or $2^3 \times 3$

Exercises

Determine whether each number is prime or composite.

1. 27   2. 31   3. 46   4. 53

5. 11   6. 72   7. 17   8. 51

Determine the prime factorization of the following numbers.

9. 64   10. 100   11. 45   12. 81

4-1 Practice

Prime Factorization

Determine whether each number is *prime* or *composite*.

1. 45	**2.** 17	**3.** 21
4. 51	**5.** 11	**6.** 71
7. 3	**8.** 27	**9.** 47

Find the prime factorization of each number.

10. 88	**11.** 39	**12.** 75
13. 124	**14.** 165	**15.** 225
16. 100	**17.** 91	**18.** 27

ALGEBRA   Factor each expression.

19. $20xy$	**20.** $18bc$	**21.** $11pqr$

22. $36g^2h^2$	**23.** $44m^2n$	**24.** $25z^2$

Replace each ● with prime factors to make a true sentence.

25. $2^2 \cdot ● \cdot 7 = 252$	**26.** $2 \cdot ● \cdot 5^3 = 750$	**27.** $2^3 \cdot ● \cdot 3^2 = 1,800$

28. ALGEBRA   Is $2x + y$ *prime* or *composite* if $x = 2$ and $y = 7$?

29. ATHLETICS   The distance around an oval running track is 440 yards. Write this distance as a product of primes.

52

4-2 Study Guide and Intervention

Greatest Common Factor

> The **greatest common factor (GCF)** of two or more numbers is the largest number that is a factor of each number. The GCF of prime numbers is 1.

Example 1 Find the GCF of 72 and 108 by listing factors.

factors of 72: 1, 2, 3, 4, 6, 8, 9, 12, 18, 24, 36, 72

factors of 108: 1, 2, 3, 4, 6, 9, 12, 18, 27, 36, 54, 108

common factors: 1, 2, 3, 4, 6, 9, 12, 18, 36

The GCF of 72 and 108 is 36.

Example 2 Find the GCF of 42 and 60 using prime factors.

Method 1 Write the prime factorization.

$60 = 2 \times 2 \times 3 \times 5$
$42 = \phantom{2 \times {}} 2 \times 3 \times 7$

Method 2 Divide by prime numbers.
Divide both 42 and 60 by 2.
Then divide the quotients by 3.

$$
\begin{array}{r}
7 \quad 10 \\
3\overline{)21 \quad 30} \\
2\overline{)42 \quad 60} \quad \leftarrow \boxed{\text{Start here}}
\end{array}
$$

The common prime factors are 2 and 3. The GCF of 42 and 60 is 2×3, or 6.

Exercises

Find the GCF of each set of numbers.

1. 18, 30

2. 60, 45

3. 24, 72

4. 32, 48

5. 100, 30

6. 54, 36

7. 3, 97, 5

8. 4, 20, 24

9. 36, 9, 45

4-2 Practice

Greatest Common Factor

Find the GCF of each set of numbers.

1. 16, 44

2. 15, 35

3. 24, 32

4. 27, 63

5. 20, 80

6. 18, 38

7. 14, 49

8. 66, 99

9. 9, 35

10. 6, 24, 42

11. 30, 50, 70

12. 32, 48, 96

13. $10w, 5w$

14. $16xy, 24xy$

15. $21ab, 35a$

16. $10jk, 15k$

17. $3mn, 9mn, 12mn$

18. $6xy, 9x, 3y$

19. 4 inches, 1 foot, 6 inches, 2 feet

20. 10 gallons, 55 gallons, 35 gallons, 20 gallons

Find two numbers whose GCF is the given number.

21. 10

22. 8

23. 14

24. SPORTS CARDS Jason wants to organize his sports cards in packets for each type of sport. Each packet has the same number of cards. If he has 24 baseball cards, 60 hockey cards, and 48 football cards, find the greatest number of cards in each packet.

25. FORESTRY A forest ranger needs to remove three tree trunks by cutting the trunks into equal lengths. If the lengths of the tree trunks are 6 feet, 8 feet, and 12 feet, what is the length of the longest log that can be cut?

4-3 Study Guide and Intervention

Problem-Solving Investigation: Make an Organized List

When solving problems often times it is useful to make an organized list. By doing so you can see all the possible solutions to the problem being posed.

Example 1 LUNCH Walnut Hills School has a deli line where students are able to select a meat sandwich, a side, and fruit. Meat choices are ham or turkey. The side choices are pretzels or chips. Fruit options are an apple or a pear. How many different combinations are possible?

Understand You know that students can choose a sandwich, a side, and fruit. There are 2 meat choices, 2 side choices, and 2 fruit choices. You need to find all possible combinations.

Plan Make an organized list.

Solve

	1	2	3	4	5	6	7	8
Meat	Ham	Ham	Ham	Ham	Turkey	Turkey	Turkey	Turkey
Side	Pretzel	Pretzel	Chips	Chips	Pretzel	Pretzel	Chips	Chips
Fruit	Apple	Pear	Apple	Pear	Apple	Pear	Apple	Pear

There are 8 possibilities.

Check Draw a tree diagram to check the result.

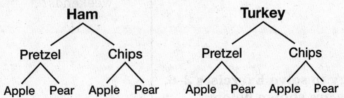

Exercises

1. Susan has 3 shirts; red, blue, and green; 2 pants; jeans and khakis; and 3 shoes; white, black, and tan, to choose from for her school outfit. How many different outfits can she create?

2. The Motor Speedway is awarding money to the first two finishers in their annual race. If there are four cars in the race numbered 1 through 4, how many different ways can they come in first and second?

Lesson 4-3

4-3 Practice

Problem-Solving Investigation: Make An Organized List

Mixed Problem Solving

For Exercises 1 and 2, solve each problem by making an organized list.

1. **VACATION** Kessler, Kacy, and their parents sit in different seats in the car while driving to their grandparents for vacation. If only the parents take turns driving, how many different ways can all four people sit in the car with 2 front and 2 back seats?

2. **PIZZA** Everyone at the table likes pepperoni, sausage, onions, and black olives on pizza. List the different possibilities of ordering a 2-topping pizza.

Use any strategy to solve Exercises 3–6. Some strategies are shown below.

PROBLEM-SOLVING STRATEGIES
• Guess and check.
• Work backward.
• Make an organized list.

3. **NUMBER SENSE** A number is increased by 12. When this sum is divided by 3, the result is the original number. What is the number?

4. **COINS** Three coins are tossed: a quarter, a nickel, and a dime. Complete the table showing the 8 different ways the coins could land by using H for heads and T for tails.

Quarter	H	H						
Nickel	H	H						
Dime	H	T						

5. **MEASUREMENT** Eight furlongs is equal to one mile. If a mile is 5,280 feet, how many feet are in 5 furlongs?

6. **TIME** Greg works at the hardware store on weekends. He worked a total of 53 hours during the month of April. How many hours did Greg work during the last weekend in April, if he worked 14 hours, 12 hours, and 15 hours the other weekends?

4-4 Study Guide and Intervention

Simplifying Fractions

Fractions that have the same value are called **equivalent fractions**. A fraction is in **simplest form** when the GCF of the numerator and denominator is 1.

Example 1 Write $\frac{36}{54}$ in simplest form.

First, find the GCF of the numerator and denominator.
factors of 36: 1, 2, 3, 4, 6, 9, 12, 18, 36
factors of 54: 1, 2, 3, 6, 9, 18, 27, 54
The GCF of 36 and 54 is 18.

Then, divide the numerator and the denominator by the GCF.
$\frac{36}{54} = \frac{36 \div 18}{54 \div 18} = \frac{2}{3}$ So, $\frac{36}{54}$ written in simplest form is $\frac{2}{3}$.

Example 2 Write $\frac{8}{12}$ in simplest form.

Find the GCF of the numerator and the denominator.

factors of $8 = 2 \cdot 2 \cdot 2$

factors of $12 = 2 \cdot 2 \cdot 3$

The GCF of 8 and 12 is $2 \cdot 2$ or 4.

$\frac{8 \div 4}{12 \div 4} = \frac{2}{3}$

So, $\frac{8}{12}$ written in simplest form is $\frac{2}{3}$.

$\frac{8}{12}$ = $\frac{2}{3}$

Exercises

Write each fraction in simplest form.

1. $\frac{42}{72}$

2. $\frac{40}{64}$

3. $\frac{21}{35}$

4. $\frac{25}{100}$

5. $\frac{99}{132}$

6. $\frac{17}{85}$

Lesson 4-4

4-4 Practice

Simplifying Fractions

Write each fraction in simplest form.

1. $\dfrac{12}{15}$

2. $\dfrac{20}{45}$

3. $\dfrac{8}{24}$

4. $\dfrac{22}{30}$

5. $\dfrac{30}{90}$

6. $\dfrac{29}{29}$

7. $\dfrac{77}{88}$

8. $\dfrac{32}{48}$

9. $\dfrac{21}{35}$

10. $\dfrac{63}{99}$

11. $\dfrac{18}{36}$

12. $\dfrac{24}{30}$

13. $\dfrac{30}{75}$

14. $\dfrac{12}{60}$

15. $\dfrac{16}{36}$

16. $\dfrac{42}{49}$

17. $\dfrac{55}{100}$

18. $\dfrac{150}{180}$

19. $\dfrac{35}{140}$

20. $\dfrac{90}{135}$

21. **STATES** Eight states in the United States start with the letter M. What fraction of states, in simplest form, begins with the letter M?

22. **MEASUREMENT** Fifteen inches is what fraction, in simplest form, of a yard?

23. **PERIMETER** A rectangle has length 7 centimeters and width 4 centimeters. What fraction of the perimeter, in simplest form, is the width?

24. **MONEY** Thirty-five cents is what fraction, in simplest form, of a dollar?

25. **AGE** Angie is 6 years old. Her dad is 30 years old. Angie's age is what fraction, in simplest form, of her dad's age?

4-5 Study Guide and Intervention

Fractions and Decimals

To write a decimal as a fraction, divide the numerator of the fraction by the denominator. Use a power of ten to change a decimal to a fraction.

Example 1 Write $\frac{5}{9}$ as a decimal.

Method 1 Use pencil and paper.

$$
\begin{array}{r}
0.555... \\
9\overline{)5.000} \\
4\,5 \\
\hline
50 \\
45 \\
\hline
50 \\
45 \\
\hline
5
\end{array}
$$

The remainder after each step is 5.

Method 2 Use a calculator.

5 ÷ 9 = 0.55555556

You can use bar notation $0.\overline{5}$ to indicate that 5 repeats forever. So, $\frac{5}{9} = 0.\overline{5}$.

Example 2 Write 0.32 as a fraction in simplest form.

$0.32 = \frac{32}{100}$ The 2 is in the hundredths place.

$\quad = \frac{8}{25}$ Simplify.

Exercises

Write each fraction or mixed number as a decimal. Use bar notation if the decimal is a repeating decimal.

1. $\frac{8}{10}$

2. $\frac{3}{5}$

3. $\frac{7}{11}$

4. $4\frac{7}{8}$

5. $\frac{13}{15}$

6. $3\frac{47}{99}$

Write each decimal as a fraction in simplest form.

7. 0.14

8. 0.3

9. 0.94

Lesson 4-5

4-5 Practice

Fractions and Decimals

Write each fraction or mixed number as a decimal. Use bar notation if the decimal is a repeating decimal.

1. $\dfrac{5}{8}$ 2. $\dfrac{2}{9}$ 3. $\dfrac{37}{16}$

4. $\dfrac{3}{4}$ 5. $\dfrac{27}{50}$ 6. $\dfrac{121}{25}$

7. $\dfrac{5}{6}$ 8. $\dfrac{1}{33}$ 9. $\dfrac{62}{11}$

10. $\dfrac{2}{3}$ 11. $\dfrac{11}{40}$ 12. $\dfrac{13}{20}$

13. $\dfrac{83}{5}$ 14. $\dfrac{3}{10}$ 15. $\dfrac{1}{9}$

16. $\dfrac{3}{7}$ 17. $\dfrac{111}{24}$ 18. $\dfrac{7}{32}$

Write each decimal as a fraction or mixed number in simplest form.

19. 0.4   20. 0.83   21. 3.75

22. 2.42   23. 0.16   24. 0.65

25. **KILOMETERS**   One kilometer is approximately 0.62 mile. What fraction represents this length?

26. **MARATHON**   Jake completed a marathon race in 3 hours and 12 minutes. Write Jake's running time as a decimal.

4-6 Study Guide and Intervention

Fractions and Percents

A **ratio** is a comparison of two numbers by division. When a ratio compares a number to 100, it can be written as a **percent**. To write a ratio or fraction as a percent, find an equivalent fraction with a denominator of 100. You can also use the meaning of percent to change percents to fractions.

Example 1 Write $\frac{19}{20}$ as a percent.

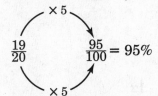

$\frac{19}{20}$ → $\frac{95}{100} = 95\%$ Since $100 \div 20 = 5$, multiply the numerator and denominator by 5.

Example 2 Write 92% as a fraction in simplest form.

$92\% = \frac{92}{100}$ Definition of percent

$\quad\ \ = \frac{23}{25}$ Simplify.

Exercises

Write each ratio as a percent.

1. $\frac{14}{100}$

2. $\frac{27}{100}$

3. 34.5 per 100

4. 18 per 100

5. 21:100

6. 96:100

Write each fraction as a percent.

7. $\frac{3}{100}$

8. $\frac{14}{100}$

9. $\frac{2}{5}$

10. $\frac{1}{20}$

11. $\frac{13}{25}$

12. $\frac{4}{10}$

Write each percent as a fraction in simplest form.

13. 35%

14. 18%

15. 75%

16. 80%

17. 16%

18. 15%

4-6 **Practice**

Fractions and Percents

Write each ratio as a percent.

1. 56 out of 100 CDs sold

2. 75 per 100 adults

3. 89.2 out of 100 hours worked

4. 26.5:100 Calories

5. $45\frac{7}{8}$ out of 100 meters

6. $33\frac{1}{3}$:100 minutes

Write each fraction as a percent.

7. $\frac{6}{10}$

8. $\frac{7}{20}$

9. $\frac{21}{25}$

10. $\frac{12}{50}$

11. $\frac{1}{2}$

12. $\frac{4}{5}$

13. $\frac{20}{90}$

14. $\frac{24}{25}$

Write each percent as a fraction in simplest form.

15. 40%

16. 35%

17. 72%

18. 44%

19. 90%

20. 17%

21. 5%

22. 26%

Replace each ● with >, <, or = to make a true sentence.

23. $\frac{1}{10}$ ● 15%

24. $\frac{3}{4}$ ● 72%

25. 85% ● $\frac{17}{20}$

26. $\frac{21}{25}$ ● 21%

27. 27% ● $\frac{27}{50}$

28. $\frac{4}{5}$ ● 60%

29. SPORTS If twenty-seven out of every 50 sports fans attend at least one professional game every year, what percent of sports fans attend at least one professional game every year?

30. WEATHER It rained 18 days during the month of April. What percent of the days during the month of April did it not rain?

4-7 Study Guide and Intervention

Percents and Decimals

To write a percent as a decimal, divide the percent by 100 and remove the percent symbol. To write a decimal as a percent, multiply the decimal by 100 and add the percent symbol.

Example 1 Write 42.5% as a decimal.

$42.5\% = \dfrac{42.5}{100}$ Write the percent as a fraction.

$\quad = \dfrac{42.5 \times 10}{100 \times 10}$ Multiply by 10 to remove the decimal in the numerator.

$\quad = \dfrac{425}{1,000}$ Simplify.

$\quad = 0.425$ Write the fraction as a decimal.

Example 2 Write 0.625 as a percent.

$0.625 = 062.5$ Multiply by 100.

$\quad = 62.5\%$ Add the % symbol.

Exercises

Write each percent as a decimal.

1. 6%

2. 28%

3. 81%

4. 84%

5. 35.5%

6. 12.5%

7. 14.2%

8. 11.1%

Write each decimal as a percent.

9. 0.47

10. 0.03

11. 0.075

12. 0.914

4-7 Practice

Percents and Decimals

Write each percent as a decimal.

1. 35%   2. 90%   3. 5 %   4. 1%

5. 21.8%   6. 64.8%   7. 4.1%   8. 8.5%

9. $39\frac{21}{50}\%$ 10. $17\frac{2}{5}\%$ 11. $40\frac{3}{4}\%$ 12. $88\frac{3}{5}\%$

Write each decimal as a percent.

13. 0.4   14. 0.8   15. 3.7   16. 9.1

17. 0.77   18. 0.03   19. 0.25   20. 0.59

21. 0.375   22. 0.123   23. 0.005   24. 0.6019

Replace each ● with >, <, or = to make a true sentence.

25. 1.5 ● 15%   26. 0.88 ● 8.8%   27. 33% ● 0.33

28. 90% ● 0.09   29. 0.26 ● 27%   30. 65.4% ● 0.645

ANALYZE TABLES For Exercises 31–33, use the table and the information given.

The table lists the approximate milk fat content of 5 types of milk products.

31. Which product has the highest milk fat content?

32. Find the approximate number of grams of milk fat in a 200-gram serving of whole milk.

33. Which milk product will have approximately 15.36 grams of milk fat in an 80-gram serving?

Milk Product	Percent Milk Fat
Heavy Cream	36.7%
Light Cream	19.2%
Whole Milk	3.5%
Low-Fat Milk	1.5%
Skim Milk	0.05%

4-8 Study Guide and Intervention

Least Common Multiple

A **multiple** of a number is the product of that number and any whole number. The least nonzero multiple of two or more numbers is the **least common multiple (LCM)** of the numbers.

Example 1 Find the LCM of 15 and 20 by listing multiples.

List the multiples.

multiples of 15: 15, 30, 45, **60**, 75, 90, 105, **120**, ...

multiples of 20: 20, 40, **60**, 80, 100, **120**, 140, ...

Notice that 60, 120, ..., are common multiples. So, the LCM of 15 and 20 is 60.

Example 2 Find the LCM of 8 and 12 using prime factors.

Write the prime factorization.

$8 = 2 \times 2 \times 2 = 2^3$
$12 = 2 \times 2 \times 3 = 2^2 \times 3$

The prime factors of 8 and 12 are 2 and 3.
Multiply the greatest power of both 2 and 3.

The LCM of 8 and 12 is $2^3 \times 3$, or 24.

Exercises

Find the LCM of each set of numbers.

1. 4, 6

2. 6, 9

3. 5, 9

4. 8, 10

5. 12, 15

6. 15, 21

7. 4, 15

8. 8, 20

9. 8, 16

10. 6, 14

11. 12, 20

12. 9, 12

13. 14, 21

14. 6, 15

15. 4, 6, 8

16. 3, 5, 6

Lesson 4-8

4-8 Practice

Least Common Multiple

Find the LCM of each set of numbers.

1. 8, 12   **2.** 10, 25   **3.** 12, 18

4. 20, 30   **5.** 8, 9   **6.** 15, 35

7. 3, 5, 7   **8.** 4, 10, 12   **9.** 9, 12, 15

10. 5, 15, 20   **11.** 14, 21, 42   **12.** 15, 18, 30

13. 2 feet, 1 yard   **14.** 6¢, 18¢, 24¢   **15.** 40 seconds, 1 minute

Write two numbers whose LCM is the given number.

16. 24   **17.** 63   **18.** 50

19. SECURITY In a large industrial complex, three security teams work different types of security checks. The first team makes a complete round in 3 hours, the second team makes a complete round in 2 hours, while the third team makes a complete round in 4 hours. If all three teams start security checks at 7 A.M., when will be the next time all three teams finish a security check at the same time?

20. COOKIES A recipe for large oatmeal cookies will make 15 cookies. A recipe for chocolate chip cookies will make 2 dozen cookies. If you want to have the same number of each type of cookie, what is the least number of each that you will need to make using complete recipes?

21. ICE SKATING Three friends ice skate at different speeds. Parcel skates one lap in 45 seconds. It takes Hansel $1\frac{1}{2}$ minutes to skate one lap and Forrest takes only 30 seconds to skate a lap. If they started out together, in how many minutes will they meet next?

4-9 Study Guide and Intervention

Comparing and Ordering Rational Numbers

To compare fractions, rewrite them so they have the same denominator. The **least common denominator (LCD)** of two fractions is the LCM of their denominators.

Another way to compare fractions is to express them as decimals. Then compare the decimals.

Example 1 Which fraction is greater, $\frac{3}{4}$ or $\frac{4}{5}$?

Method 1 Rename using the LCD.

$$\frac{3}{4} = \frac{3 \times 5}{4 \times 5} = \frac{15}{20}$$

$$\frac{4}{5} = \frac{4 \times 4}{5 \times 4} = \frac{16}{20}$$

The LCD is 20.

Because the denominators are the same, compare numerators.

Since $\frac{16}{20} > \frac{15}{20}$, then $\frac{4}{5} > \frac{3}{4}$.

Method 2 Write each fraction as a decimal. Then compare decimals.

$$\frac{3}{4} = 0.75$$

$$\frac{4}{5} = 0.8$$

Since $0.8 > 0.75$, then $\frac{4}{5} > \frac{3}{4}$.

Exercises

Find the LCD of each pair of fractions.

1. $\frac{1}{2}, \frac{1}{8}$

2. $\frac{1}{3}, \frac{3}{4}$

3. $\frac{3}{4}, \frac{7}{10}$

Replace each ● with <, >, or = to make a true sentence.

4. $\frac{1}{2} \bullet \frac{4}{9}$

5. $\frac{4}{5} \bullet \frac{8}{10}$

6. $\frac{3}{4} \bullet \frac{7}{8}$

7. $\frac{1}{2} \bullet \frac{5}{9}$

8. $\frac{9}{14} \bullet \frac{10}{17}$

9. $\frac{5}{7} \bullet \frac{6}{11}$

10. $\frac{8}{17} \bullet \frac{1}{2}$

11. $\frac{9}{10} \bullet \frac{17}{19}$

4-9 Practice

Comparing and Ordering Rational Numbers

Replace each ● with >, <, or = to make a true sentence.

1. $\dfrac{5}{6}$ ● $\dfrac{1}{3}$

2. $\dfrac{4}{5}$ ● $\dfrac{9}{10}$

3. $\dfrac{6}{9}$ ● $\dfrac{4}{6}$

4. $\dfrac{2}{7}$ ● $\dfrac{1}{8}$

5. $\dfrac{15}{21}$ ● $\dfrac{12}{18}$

6. $\dfrac{24}{32}$ ● $\dfrac{36}{48}$

7. $\dfrac{8}{11}$ ● $\dfrac{10}{13}$

8. $\dfrac{14}{15}$ ● $\dfrac{19}{20}$

9. $4\dfrac{1}{5}$ ● $4\dfrac{2}{10}$

10. $7\dfrac{4}{9}$ ● $7\dfrac{2}{3}$

11. $1\dfrac{17}{20}$ ● $1\dfrac{8}{10}$

12. $9\dfrac{3}{2}$ ● $9\dfrac{5}{6}$

13. 50% ● 8 out of 10

14. 0.65 ● 65 out of 100

15. 4 out of 5 ● 75%

16. 1 out of 3 ● 1.3

17. $\dfrac{2}{3}$ mile ● $\dfrac{2}{5}$ mile

18. $\dfrac{7}{10}$ gram ● 0.72 gram

Determine whether each number is rational. Write *yes* or *no*. Explain your reasoning.

19. $\dfrac{8}{21}$

20. 0.50550555 . . .

21. $1.\overline{142857}$

Order each set of numbers from least to greatest.

22. 63%, $\dfrac{2}{3}$, 0.65

23. $\dfrac{7}{8}$, 0.98, 98.5%,

24. 0.2, 2%, $\dfrac{1}{12}$

25. **BASEBALL** The pitchers for the home team had 12 strikeouts for 32 batters, while the pitchers for the visiting team had 15 strikeouts for 35 batters. Which pitching team had a greater fraction of strikeouts?

26. **TRANSPORTATION** To get to school, 38% of the students ride in the family vehicle, 5 out of 12 students ride on the school bus, and 0.12 of the students ride a bike. Order the types of transportation students use to get to school from least to greatest.

5-1 Study Guide and Intervention

Estimating with Fractions

Use rounding to estimate with fractions.

Estimating:	For mixed numbers, round to the nearest whole number.	For fractions, round to 0, $\frac{1}{2}$, or 1.
	$4\frac{1}{6} + 3\frac{7}{8} \rightarrow 4 + 4 = 8$	$\frac{11}{12} - \frac{4}{9} \rightarrow 1 - \frac{1}{2} = \frac{1}{2}$
	$4\frac{1}{6} + 3\frac{7}{8}$ is *about* 8.	$\frac{11}{12} - \frac{4}{9}$ is *about* $\frac{1}{2}$.

Example 1 Estimate $2\frac{2}{3} \times 4\frac{1}{4}$.

$2\frac{2}{3} \times 4\frac{1}{4} \rightarrow 3 \times 4 = 12$

The product is *about* 12.

Example 2 Estimate $\frac{6}{7} - \frac{3}{5}$.

$\frac{6}{7}$ is about 1.

$\frac{3}{5}$ is about $\frac{1}{2}$.

$\frac{6}{7} - \frac{3}{5} \rightarrow 1 - \frac{1}{2} = \frac{1}{2}$ The difference is *about* $\frac{1}{2}$.

Exercises

Estimate.

1. $4\frac{1}{3} + 3\frac{4}{5}$

2. $2\frac{1}{6} \times 3\frac{2}{3}$

3. $\frac{7}{12} - \frac{1}{10}$

4. $5\frac{1}{4} - 1\frac{1}{2}$

5. $4\frac{3}{4} + 1\frac{1}{5}$

6. $\frac{5}{9} \times \frac{13}{14}$

7. $\frac{1}{6} \div \frac{8}{9}$

8. $\frac{6}{7} \div \frac{9}{10}$

9. $13\frac{4}{5} \div 1\frac{7}{8}$

10. $12\frac{1}{4} \div 5\frac{7}{8}$

Lesson 5-1

5-1 Practice

Estimating with Fractions

Estimate.

1. $7\frac{1}{6} + 5\frac{8}{9}$

2. $4\frac{2}{10} + 1\frac{1}{2}$

3. $\frac{11}{13} - \frac{15}{16}$

4. $6\frac{4}{5} \cdot 3\frac{2}{7}$

5. $\frac{6}{11} - \frac{1}{5}$

6. $8\frac{1}{4} \div 3\frac{7}{8}$

7. $\frac{1}{8} \div \frac{17}{20}$

8. $\frac{5}{8} \cdot \frac{9}{10}$

9. $9\frac{14}{15} - 2\frac{3}{4}$

10. $5\frac{3}{5} \div \frac{5}{6}$

11. $\frac{10}{11} \cdot 1\frac{1}{9}$

12. $4\frac{1}{14} + 5\frac{7}{8}$

13. $5\frac{1}{9} + 1\frac{6}{7} + \frac{5}{6}$

14. $4\frac{9}{10}\left(2\frac{1}{3} + \frac{7}{8}\right)$

15. $3\frac{1}{5}\left(7\frac{2}{3} - 1\frac{8}{9}\right)$

Estimate using compatible numbers.

16. $\frac{1}{5} \cdot 44$

17. $\frac{1}{7} \cdot 29$

18. $33\frac{1}{10} \div 4\frac{1}{3}$

19. $\frac{1}{8} \cdot 62$

20. $20\frac{5}{6} \div 6\frac{2}{5}$

21. $19\frac{4}{5} \div 8\frac{2}{3}$

ANALYZE TABLES For Exercises 22–24, use the following information and the table shown.

For a recent year, the table shows the approximate number of dollars spent in each category by consumers in Kansas City for every $100 spent.

22. About how many dollars are spent on apparel and entertainment for every $100 spent?

Expenditure	Dollars Spent for Every $100 Spent
Apparel	$3\frac{7}{10}$
Health Care	$5\frac{3}{5}$
Entertainment	$5\frac{3}{10}$

23. What is the approximate difference in spending for health care and entertainment for every $100 spent?

24. What is the approximate amount of money spent for all three areas for every $100 spent?

5-2 Study Guide and Intervention

Adding and Subtracting Fractions

Like fractions are fractions that have the same denominator. To add or subtract like fractions, add or subtract the numerators and write the result over the denominator.

Simplify if necessary.

To add or subtract *unlike fractions*, rename the fractions with a least common denominator. Then add or subtract as with like fractions.

Example 1 Subtract $\frac{3}{4} - \frac{1}{4}$. Write in simplest form.

$$\frac{3}{4} - \frac{1}{4} = \frac{3-1}{4}$$ Subtract the numerators.

$$= \frac{2}{4}$$ Write the difference over the denominator.

$$= \frac{1}{2}$$ Simplify.

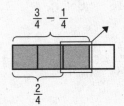

Example 2 Add $\frac{2}{3} + \frac{1}{12}$. Write in simplest form.

The least common denominator of 3 and 12 is 12.

$$\frac{2}{3} = \frac{2 \times 4}{3 \times 4} = \frac{8}{12}$$ Rename $\frac{2}{3}$ using the LCD.

$$\begin{array}{ccc} \frac{2}{3} & \rightarrow & \frac{8}{12} \\ +\frac{1}{12} & \rightarrow & +\frac{1}{12} \\ \hline & & \frac{9}{12} \text{ or } \frac{3}{4} \end{array}$$ Add the numerators and simplify.

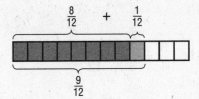

Exercises

Add or subtract. Write in simplest form.

1. $\frac{5}{8} + \frac{1}{8}$ 2. $\frac{7}{9} - \frac{2}{9}$

3. $\frac{1}{2} + \frac{3}{4}$ 4. $\frac{7}{8} - \frac{5}{6}$

5. $\frac{5}{9} + \frac{5}{6}$ 6. $\frac{3}{8} - \frac{1}{12}$

7. $\frac{3}{10} + \frac{7}{12}$ 8. $\frac{2}{5} - \frac{1}{3}$

9. $\frac{7}{15} + \frac{5}{6}$ 10. $\frac{7}{9} - \frac{1}{2}$

Lesson 5-2

5-2 Practice

Adding and Subtracting Fractions

Add or subtract. Write in simplest form.

1. $\dfrac{2}{5} + \dfrac{3}{5}$

2. $\dfrac{2}{9} + \dfrac{4}{9}$

3. $\dfrac{8}{11} - \dfrac{7}{11}$

4. $\dfrac{4}{8} + \dfrac{5}{8}$

5. $\dfrac{1}{18} + \dfrac{5}{6}$

6. $\dfrac{7}{15} - \dfrac{1}{5}$

7. $\dfrac{9}{16} - \dfrac{5}{12}$

8. $\dfrac{5}{14} - \dfrac{2}{21}$

9. $\dfrac{7}{8} - \dfrac{1}{6}$

10. $\dfrac{7}{10} - \dfrac{4}{15}$

11. $\dfrac{5}{6} - \dfrac{3}{4}$

12. $\dfrac{2}{3} - \dfrac{1}{2}$

13. $1 + \dfrac{1}{6}$

14. $1 - \dfrac{3}{5}$

15. $4 + \dfrac{8}{9}$

16. $5 - \dfrac{1}{4}$

17. $\dfrac{2}{3} + \dfrac{4}{15} + \dfrac{1}{5}$

18. $\dfrac{7}{8} + \dfrac{1}{2} + \dfrac{3}{16}$

19. $\left(\dfrac{3}{4} + \dfrac{1}{3}\right) - \dfrac{11}{12}$

20. $\left(\dfrac{4}{5} - \dfrac{7}{10}\right) + \dfrac{1}{4}$

21. **STATES** Most of the state names in the United States end in a vowel. Of the 50 states, $\dfrac{1}{2}$ of the state names end in either an a or an e and $\dfrac{3}{25}$ end in either an i or an o. If none of the state names end in a u, what is the fraction of state names that end in a vowel?

22. **JIGSAW PUZZLES** Over the weekend, Halverson had put together $\dfrac{3}{16}$ of a jigsaw puzzle, while Jaime put together $\dfrac{5}{8}$ of the puzzle. Who had completed a greater fraction of the jigsaw puzzle, and by how much?

ALGEBRA Evaluate each expression if $x = \dfrac{5}{8}$ and $y = \dfrac{5}{4}$.

23. $x - \dfrac{1}{2}$

24. $y - x$

25. $\dfrac{5}{16} + y$

26. $x + y$

5-3 Study Guide and Intervention

Adding and Subtracting Mixed Numbers

To add or subtract mixed numbers:

1. Add or subtract the fractions. Rename using the LCD if necessary.

2. Add or subtract the whole numbers.

3. Simplify if necessary.

Example 1 Find $14\frac{1}{2} + 18\frac{2}{3}$.

$$14\frac{1}{2} \rightarrow 14\frac{3}{6}$$ Rename the fractions.

$$+18\frac{2}{3} \rightarrow +18\frac{4}{6}$$ Add the whole numbers and add the fractions.

$$32\frac{7}{6} \text{ or } 33\frac{1}{6}$$ Simplify.

Example 2 Find $21 - 12\frac{5}{8}$.

$$21 \rightarrow 20\frac{8}{8}$$ Rename 21 as $20\frac{8}{8}$.

$$-12\frac{5}{8} \rightarrow -12\frac{5}{8}$$ First subtract the whole numbers and then the fractions.

$$8\frac{3}{8}$$

Exercises

Add or subtract. Write in simplest form.

1. $7\frac{3}{4} + 2\frac{3}{4}$ 2. $14\frac{2}{9} - 6\frac{1}{9}$ 3. $9\frac{1}{5} - 4\frac{3}{4}$

4. $7\frac{1}{8} + 5\frac{3}{8}$ 5. $7\frac{3}{4} + 2\frac{2}{3}$ 6. $5\frac{1}{2} - 5\frac{1}{3}$

7. $5\frac{1}{2} - 3\frac{1}{4}$ 8. $6\frac{1}{3} + 2\frac{1}{6}$ 9. $9 - 3\frac{2}{5}$

10. $2\frac{2}{3} + 7\frac{1}{2}$ 11. $6\frac{1}{2} - 6\frac{1}{3}$ 12. $18\frac{1}{2} + 5\frac{5}{8}$

5-3 Practice

Adding and Subtracting Mixed Numbers

Add or Subtract. Write in simplest form.

1. $3\frac{1}{8} + 5\frac{3}{8}$ 2. $4\frac{1}{6} + 7\frac{1}{6}$ 3. $9\frac{3}{4} - 6\frac{1}{4}$ 4. $5\frac{5}{9} - 4\frac{2}{9}$

5. $8\frac{2}{3} - 3\frac{1}{6}$ 6. $10\frac{3}{4} - 5\frac{3}{8}$ 7. $7\frac{3}{10} + 12\frac{2}{5}$ 8. $1\frac{1}{6} + 1\frac{1}{8}$

9. $5\frac{1}{3} - 3\frac{2}{3}$ 10. $8\frac{4}{7} - 7\frac{5}{7}$ 11. $11\frac{1}{12} - 6\frac{5}{6}$ 12. $3\frac{2}{5} - 1\frac{3}{4}$

13. $5\frac{4}{5} + 6\frac{5}{6}$ 14. $8\frac{2}{7} + 6\frac{5}{14}$ 15. $9 - 7\frac{3}{8}$ 16. $6\frac{4}{5} + 7\frac{1}{5}$

17. $4\frac{3}{5} + 1\frac{11}{20} + 5\frac{7}{10}$ 18. $10 - 9\frac{1}{3}$ 19. $2\frac{1}{4} + 5\frac{3}{8} + 3\frac{1}{2}$ 20. $7 - 6\frac{7}{8}$

21. **LAND MEASUREMENT**   Mr. Alfonso owns two adjacent pieces of land totaling $13\frac{3}{8}$ acres. One piece of land is $8\frac{7}{12}$ acres. Find the area of the other piece of land.

GEOMETRY   Find the perimeter of each figure.

22.

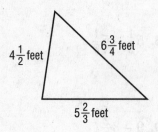

23.

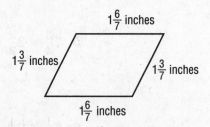

5-4 Study Guide and Intervention

Problem-Solving Investigation: Eliminate Possibilities

By **eliminating possibilities** when problem solving, you can methodically reduce the number of potential answers.

Example

Joan has $20 to spend on her sister for her birthday. She has already bought her a DVD for $9.75. There are three shirts that she likes which cost $8.75, $10.00, and $11.00. Which shirt should she buy so that she spends most of her money without going over $20?

Understand	You know that the total amount of money she has to spend must be $20 or less.
Plan	Eliminate answers that are not reasonable.
Solve	She couldn't spend $11.00 because $9.75 + $11.00 = $20.75. So eliminate that choice. Now check $10.00 $9.75 + 10.00 = $19.75 Since this is less than $20, this is the correct choice. She should buy her sister the $10.00 shirt.
Check	By buying the $8.75 shirt, she would only spend a total of $9.75 + $8.75 = $18.50. This is less than the $20 minimum, but not the most she could possibly spend.

Exercises

Solve the following problems by eliminating possibilities first.

1. **TELEPHONE** Susan talked on her cellular telephone for 120 minutes last month. Her plan charges her a $15.00 fee per month plus $0.10 a minute after the first 60 minutes, which are included in the $15 fee. What was her total bill for last month?

   A. $12.00   B. $27.00   C. $21.00   D. $6.00

2. **HOME SALES** 450 homes sold in your area in the last year. What number shows a good estimate of the number of homes sold per month?

   A. 38 homes   B. 32.5 homes   C. 2 homes   D. 45 homes

3. **CAR SALES** Derrick sells cars for a living. He sells an average of 22 cars a month. What will his total average car sales be in 5 years?

   A. 110 cars   B. 264 cars   C. 1320 cars   D. 27 cars

4. **TELEVISION** Myra is allowed to watch 6 hours of television on a weekend. She watched $2\frac{1}{2}$ hours this morning. How much television will she be allowed to watch at most this afternoon?

   A. 4 hours   B. $4\frac{1}{2}$ hours   C. $2\frac{1}{2}$ hours   D. $3\frac{1}{2}$ hours

Lesson 5-4

5-4 Practice

Problem-Solving Investigation: Eliminate Possibilities

Mixed Problem Solving

Eliminate the possibilities to solve
Exercises 1 and 2.

1. **STAIRCASE** A staircase has 14 steps
between floors. If the second floor is
10 feet above the first floor, what is the
approximate height of each step of the
staircase?

   A 2 inches   C 9 inches

   B 12 inches   D 15 inches

2. **NEWSPAPER** Mr. Kemper delivers the
morning newspaper to about 500
customers each day. About how many
newspapers does he deliver in a month?

   F 50   H 500

   G 5,000   J 15,000

Use any strategy to solve Exercises 3
and 4. Some strategies are shown below.

PROBLEM-SOLVING STRATEGIES
• Look for a pattern.
• Choose the method of computation.
• Eliminate possibilities.

3. **PATTERNS** What are the next three
fractions in the pattern?
$$\frac{1}{12}, \frac{1}{6}, \frac{1}{4}, \frac{1}{3}, \frac{5}{12}, \cdots$$

4. **OFFICE SUPPLIES** Printer ink costs $23.42
per cartridge if bought separately. If
bought by the case of 24 cartridges, the
cost per cartridge is only $19.53. About
how much is the difference in cost of
buying 4 cases than buying the same
number of cartridges separately?

   A $4   C $84

   B $374   D $1,550

Select the Operation

For Exercises 5 and 6, select the
appropriate operation(s) to solve the
problem. Justify your selection(s) and
solve the problem.

5. **YARDWORK** David mowed $\frac{3}{10}$ of the
yard while his brother mowed $\frac{1}{4}$ of it.
What fraction of the yard still needs to
be mowed?

6. **DOGS** On average, dogs require about
35 Calories per pound of body weight
per day. The Parkers own three dogs
that weigh 22 pounds, 34 pounds, and 9
pounds. What is the total Calorie
requirement for the dogs each day?

5-5 Study Guide and Intervention

Multiplying Fractions and Mixed Numbers

To multiply fractions, multiply the numerators and multiply the denominators.

$$\frac{5}{6} \times \frac{3}{5} = \frac{5 \times 3}{6 \times 5} = \frac{15}{30} = \frac{1}{2}$$

To multiply mixed numbers, rename each mixed number as a fraction. Then multiply the fractions.

$$2\frac{2}{3} \times 1\frac{1}{4} = \frac{8}{3} \times \frac{5}{4} = \frac{40}{12} = 3\frac{1}{3}$$

Example 1 Find $\frac{2}{3} \times \frac{4}{5}$. Write in simplest form.

$\frac{2}{3} \times \frac{4}{5} = \frac{2 \times 4}{3 \times 5}$ ← Multiply the numerators.
                 ← Multiply the denominators.

$\quad = \frac{8}{15}$ Simplify.

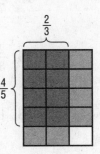

Example 2 Find $\frac{1}{3} \times 2\frac{1}{2}$. Write in simplest form.

$\frac{1}{3} \times 2\frac{1}{2} = \frac{1}{3} \times \frac{5}{2}$ Rename $2\frac{1}{2}$ as an improper fraction, $\frac{5}{2}$.

$\quad = \frac{1 \times 5}{3 \times 2}$ Multiply.

$\quad = \frac{5}{6}$ Simplify.

Exercises

Multiply. Write in simplest form.

1. $\frac{2}{3} \times \frac{2}{3}$

2. $\frac{1}{2} \times \frac{7}{8}$

3. $\frac{1}{3} \times \frac{3}{5}$

4. $\frac{5}{9} \times 4$

5. $1\frac{2}{3} \times \frac{3}{5}$

6. $3\frac{3}{4} \times 1\frac{1}{6}$

7. $\frac{3}{4} \times 1\frac{2}{3}$

8. $3\frac{1}{3} \times 2\frac{1}{2}$

9. $4\frac{1}{5} \times \frac{1}{7}$

10. $\frac{7}{5} \times 8$

11. $2\frac{1}{3} \times \frac{4}{6}$

12. $\frac{1}{8} \times 2\frac{3}{4}$

5-5 Practice

Multiplying Fractions and Mixed Numbers

Multiply. Write in simplest form.

1. $\dfrac{3}{5} \times \dfrac{1}{2}$

2. $\dfrac{3}{4} \times \dfrac{2}{7}$

3. $10 \times \dfrac{1}{3}$

4. $\dfrac{5}{8} \times 7$

5. $\dfrac{1}{7} \times \dfrac{7}{9}$

6. $\dfrac{6}{11} \times \dfrac{1}{6}$

7. $\dfrac{5}{6} \times \dfrac{1}{5}$

8. $\dfrac{1}{8} \times \dfrac{4}{5}$

9. $\dfrac{3}{8} \times \dfrac{8}{9}$

10. $\dfrac{4}{7} \times \dfrac{21}{32}$

11. $\dfrac{5}{8} \times \dfrac{18}{25}$

12. $\dfrac{20}{21} \times \dfrac{3}{5}$

13. $3\dfrac{1}{5} \times \dfrac{3}{8}$

14. $\dfrac{2}{3} \times 4\dfrac{1}{3}$

15. $15 \times 2\dfrac{2}{5}$

16. $5\dfrac{1}{2} \times 4$

17. $8 \times 3\dfrac{3}{8}$

18. $10 \times 1\dfrac{1}{15}$

19. $5\dfrac{1}{4} \times 4\dfrac{2}{3}$

20. $2\dfrac{2}{7} \times 1\dfrac{1}{8}$

For Exercises 21 and 22, use measurement conversions.

21. Find $\dfrac{1}{10}$ of $\dfrac{1}{100}$ of a meter.

22. Find $\dfrac{1}{60}$ of $\dfrac{1}{60}$ of an hour.

For Exercises 23–25, evaluate each verbal expression.

23. one-fourth of two-thirds   24. three-fifths of one-sixth   25. two-fifths of one-half

26. **GASOLINE** Jamal filled his gas tank and then used $\dfrac{7}{16}$ of the tank for traveling to visit his grandfather. He then used $\dfrac{1}{3}$ of the remaining gas in the tank to run errands around town. What fraction of the tank is filled with gasoline?

27. **HIKING** A hiker averages $6\dfrac{3}{8}$ kilometers per hour. If he hikes for $5\dfrac{1}{3}$ hours, how many kilometers did he hike?

ALGEBRA Evaluate each expression if $x = 3\dfrac{1}{3}$, $y = 4\dfrac{5}{6}$, and $z = 2$.

28. $x \times z - y$

29. $y \times z + x$

30. $3yz$

5-6 Study Guide and Intervention

Algebra: Solving Equations

Multiplicative inverses, or **reciprocals**, are two numbers whose product is 1. To solve an equation in which the coefficient is a fraction, multiply each side of the equation by the reciprocal of the coefficient.

Example 1 Find the multiplicative inverse of $3\frac{1}{4}$.

$$3\frac{1}{4} = \frac{13}{4} \qquad \text{Rename the mixed number as an improper fraction.}$$

$$\frac{13}{4} \cdot \frac{4}{13} = 1 \qquad \text{Multiply } \frac{13}{4} \text{ by } \frac{4}{13} \text{ to get the product 1.}$$

The multiplicative inverse of $3\frac{1}{4}$ is $\frac{4}{13}$.

Example 2 Solve $\frac{4}{5}x = 8$. Check your solution.

$$\frac{4}{5}x = 8 \qquad \text{Write the equation.}$$

$$\left(\frac{5}{4}\right)\frac{4}{5}x = \left(\frac{5}{4}\right)8 \qquad \text{Multiply each side by the reciprocal of } \frac{4}{5}, \frac{5}{4}.$$

$$x = 10 \qquad \text{Simplify.}$$

The solution is 10.

Exercises

Find the multiplicative inverse of each number.

1. $\frac{4}{9}$

2. $\frac{12}{13}$

3. $-\frac{15}{4}$

4. $6\frac{1}{7}$

Solve each equation. Check your solution.

5. $\frac{3}{5}x = 12$

6. $16 = \frac{10}{3}a$

7. $\frac{c}{2} = 7$

8. $\frac{15}{7}y = 3$

9. $\frac{m}{6} = -4$

10. $\frac{14}{3} = -\frac{7}{9}b$

Lesson 5-6

5-6 Practice

Algebra: Solving Equations

Find the multiplicative inverse of each number.

1. $\dfrac{7}{9}$

2. $\dfrac{5}{2}$

3. $\dfrac{1}{9}$

4. $\dfrac{1}{12}$

5. 4

6. 15

7. $4\dfrac{1}{3}$

8. $5\dfrac{4}{5}$

Solve each equation. Check your solution.

9. $\dfrac{a}{8} = 5$

10. $15 = \dfrac{y}{2}$

11. $\dfrac{h}{3.1} = 7$

12. $1 = \dfrac{x}{6.3}$

13. $0.9 = \dfrac{m}{2.5}$

14. $\dfrac{t}{5.4} = 9$

15. $\dfrac{3}{7}g = 9$

16. $28 = \dfrac{4}{5}d$

17. $\dfrac{3}{8}n = \dfrac{1}{4}$

18. $\dfrac{2}{5} = \dfrac{4}{5}c$

19. $\dfrac{2}{3}z = 4\dfrac{1}{4}$

20. $\dfrac{5}{6}b = 1\dfrac{7}{8}$

21. $\dfrac{p}{-4} = 7$

22. $-3 = \dfrac{w}{-5}$

23. $27.3 = \dfrac{3}{4}y$

24. $\dfrac{4}{7}x = -1.6$

25. **DRAWING** An architect needs to make a scale drawing of a home. The width w of the home in the drawing, in inches, is given by the equation $\dfrac{w}{0.6} = 9.5$. What is the width of the home in the scale drawing?

26. **VOLUNTEERS** At a local shelter, 36 people volunteered to help prepare meals for disaster victims. If this represented $\dfrac{9}{16}$ of the volunteers at the shelter, write and solve an equation to determine how many volunteers helped at the local shelter.

5-7 Study Guide and Intervention

Dividing Fractions and Mixed Numbers

To divide by a fraction, multiply by its multiplicative inverse or reciprocal. To divide by a mixed number, rename the mixed number as an improper fraction.

Example 1 Find $3\frac{1}{3} \div \frac{2}{9}$. Write in simplest form.

$3\frac{1}{3} \div \frac{2}{9} = \frac{10}{3} \div \frac{2}{9}$ Rename $3\frac{1}{3}$ as an improper fraction.

$= \frac{10}{3} \cdot \frac{9}{2}$ Multiply by the reciprocal of $\frac{2}{9}$, which is $\frac{9}{2}$.

$= \frac{\overset{5}{\cancel{10}}}{\underset{1}{\cancel{3}}} \cdot \frac{\overset{3}{\cancel{9}}}{\underset{1}{\cancel{2}}}$ Divide out common factors.

$= 15$ Multiply.

Exercises

Divide. Write in simplest form.

1. $\frac{2}{3} \div \frac{1}{4}$ 2. $\frac{2}{5} \div \frac{5}{6}$ 3. $\frac{1}{2} \div \frac{1}{5}$

4. $5 \div \frac{1}{2}$ 5. $\frac{5}{8} \div 10$ 6. $7\frac{1}{3} \div 2$

7. $\frac{5}{6} \div 3\frac{1}{2}$ 8. $36 \div 1\frac{1}{2}$ 9. $2\frac{1}{2} \div 10$

10. $5\frac{2}{5} \div 1\frac{4}{5}$ 11. $6\frac{2}{3} \div 3\frac{1}{9}$ 12. $4\frac{1}{4} \div \frac{3}{8}$

13. $4\frac{6}{7} \div 2\frac{3}{7}$ 14. $12 \div 2\frac{1}{2}$ 15. $4\frac{1}{6} \div 3\frac{1}{6}$

Lesson 5-7

5-7 Practice

Dividing Fractions and Mixed Numbers

Divide. Write in simplest form.

1. $\dfrac{3}{5} \div \dfrac{3}{4}$

2. $\dfrac{4}{7} \div \dfrac{8}{9}$

3. $\dfrac{6}{7} \div \dfrac{5}{6}$

4. $\dfrac{1}{4} \div \dfrac{1}{2}$

5. $7 \div \dfrac{1}{3}$

6. $\dfrac{6}{11} \div 2$

7. $4\dfrac{1}{5} \div 7$

8. $8 \div 4\dfrac{2}{3}$

9. $\dfrac{3}{4} \div 1\dfrac{1}{6}$

10. $\dfrac{7}{9} \div 2\dfrac{5}{8}$

11. $3\dfrac{2}{5} \div 5\dfrac{1}{10}$

12. $4\dfrac{8}{9} \div \dfrac{2}{3}$

13. $2\dfrac{3}{5} \div 1\dfrac{1}{4}$

14. $7\dfrac{1}{2} \div 2\dfrac{1}{2}$

15. $5\dfrac{1}{4} \div \dfrac{7}{8}$

16. $8\dfrac{1}{3} \div \dfrac{5}{9}$

17. **COOKING** Mrs. Lau rolls out $2\dfrac{3}{4}$ feet of dough to make noodles. If the noodles are $\dfrac{3}{8}$ of an inch wide, how many noodles will she make?

PIZZA For Exercises 18 and 19, use the table that shows the weights of three sizes of pizza.

18. How many times heavier is the extra-large pizza than the small pizza?

Pizza Size	Weight (lbs)
Extra large	$6\dfrac{1}{2}$
Medium	$3\dfrac{1}{4}$
Small	$1\dfrac{5}{8}$

19. How many times heavier is the medium pizza than the small pizza?

ALGEBRA Evaluate each expression if $a = \dfrac{2}{5}$, $b = \dfrac{3}{10}$, and $c = 2\dfrac{1}{2}$.

20. $b \div a$

21. $a \div c$

22. $3a \div b$

23. $\dfrac{1}{5}c \div a$

6-1 Study Guide and Intervention

Ratios

Any ratio can be written as a fraction. To write a ratio comparing measurements, such as units of length or units of time, both quantities must have the same unit of measure. Two ratios that have the same value are **equivalent ratios**.

Example 1 Write the ratio 15 to 9 as a fraction in simplest form.

15 to 9 $= \dfrac{15}{9}$ Write the ratio as a fraction.

$= \dfrac{5}{3}$ Simplify.

Written as a fraction in simplest form, the ratio 15 to 9 is $\dfrac{5}{3}$.

Example 2 Determine whether the ratios *10 cups of flour in 4 batches of cookies* and *15 cups of flour in 6 batches of cookies* are equivalent ratios.

Compare ratios written in simplest form.

10 cups:4 batches $= \dfrac{10 \div 2}{4 \div 2}$ or $\dfrac{5}{2}$ Divide the numerator and denominator by the GCF, 2

15 cups:6 batches $= \dfrac{15 \div 3}{6 \div 3}$ or $\dfrac{5}{2}$ Divide the numerator and denominator by the GCF, 3

Since the ratios simplify to the same fraction, the ratios of cups to batches are equivalent.

Exercises

Write each ratio as a fraction in simplest form.

1. 30 to 12

2. 5:20

3. 49:42

4. 15 to 13

5. 28 feet:35 feet

6. 24 minutes to 18 minutes

7. 75 seconds:150 seconds

8. 12 feet:60 feet

Determine whether the ratios are equivalent. Explain.

9. $\dfrac{3}{4}$ and $\dfrac{12}{16}$

10. 12:17 and 10:15

11. $\dfrac{25}{35}$ and $\dfrac{10}{14}$

12. 2 lb:36 oz and 3 lb:44 oz

13. 1 ft:4 in. and 3 ft:12 in.

Lesson 6-1

6-1 Practice

Ratios

SURVEY For Exercises 1–3, use the responses to a survey to write each ratio as a fraction in simplest form.

Survey Responses		
Yes	No	Not Sure
18	4	6

1. *yes* responses:
   no responses

2. *no* responses:
   not sure responses

3. *not sure* responses:
   total responses

COUNTY FAIR For Exercises 4–9, use the following information to write each ratio as a fraction in simplest form.

At its annual fair, Westborough County had 27 food booths and 63 game booths. A total of 1,350 adults and 3,600 children attended. The fair made a profit of $42,000. Of this money, $12,600 came from food sales.

4. adults:children

5. game booths:food booths

6. booths:profits

7. children:people

8. children:booths

9. non-food sale profits:profits

Determine whether the ratios are equivalent. Explain.

10. 18 trucks to 4 cars,
    21 trucks to 6 cars

11. $6 for every 10 people,
    $9 for every 15 people

12. 33 dinners to 6 packages,
    14 dinners to 4 packages

13. **ENGINES** A four cylinder engine produces a maximum of 110 horsepower. A six cylinder engine produces a maximum of 180 horsepower. Do these engines have an equivalent horsepower-to-cylinder ratio? Justify your answer.

ANALYZE TABLES For Exercises 14 and 15, use the information in the table that shows the crop statistics for three farms.

Farm	Acres of Soybeans	Acres of Corn
A	585	225
B	2,990	1,150
C	1,120	400

14. For which two farms is the soybeans-to-corn ratio the same? Explain.

15. Which farm has the highest soybeans-to-corn ratio? Justify your answer.

6-2 Study Guide and Intervention

Rates

A ratio that compares two quantities with different kinds of units is called a **rate**. When a rate is simplified so that it has a denominator of 1 unit, it is called a **unit rate**.

Example 1 DRIVING Alita drove her car 78 miles and used 3 gallons of gas. What is the car's gas mileage in miles per gallon?

Write the rate as a fraction. Then find an equivalent rate with a denominator of 1.

$78 \text{ miles using 3 gallons} = \dfrac{78 \text{ mi}}{3 \text{ gal}}$ Write the rate as a fraction.

$= \dfrac{78 \text{ mi} \div 3}{3 \text{ gal} \div 3}$ Divide the numerator and the denominator by 3.

$= \dfrac{26 \text{ mi}}{1 \text{ gal}}$ Simplify.

The car's gas mileage, or unit rate, is 26 miles per gallon.

Example 2 SHOPPING Joe has two different sizes of boxes of cereal from which to choose. The 12-ounce box costs $2.54, and the 18-ounce box costs $3.50. Which box costs less per ounce?

Find the unit price, or the cost per ounce, of each box. Divide the price by the number of ounces.

12-ounce box $2.54 \div 12 \text{ ounces} \approx \$0.21 \text{ per ounce}$
18-ounce box $3.50 \div 18 \text{ ounces} \approx \$0.19 \text{ per ounce}$

The 18-ounce box costs less per ounce.

Exercises

Find each unit rate. Round to the nearest hundredth if necessary.

1. 18 people in 3 vans

2. $156 for 3 books

3. 115 miles in 2 hours

4. 8 hits in 22 games

5. 65 miles in 2.7 gallons

6. 2,500 Calories in 24 hours

Choose the better unit price.

7. $12.95 for 3 pounds of nuts or $21.45 for 5 pounds of nuts

8. A 32-ounce bottle of apple juice for $2.50 or a 48-ounce bottle for $3.84.

6-2 Practice

Rates

Find each unit rate. Round to the nearest hundredth if necessary.

1. $11.49 for 3 packages

2. 2,550 gallons in 30 days

3. 88 students for 4 classes

4. 15.6 °F in 13 minutes

5. 175 Calories in 12 ounces

6. 258.5 miles in 5.5 hours

7. 549 vehicles on 9 acres

8. $920 for 40 hours

9. 13 apples for 2 pies

10. SPORTS The results of a track meet are shown. Who ran the fastest? Explain your reasoning. Round to the nearest ten thousandth.

Name	Event	Time (min)
Theo	3K Run	9.6
Esteban	5K Run	13.5
Tetsuo	10K Run	31.9

11. MANUFACTURING A machinist can produce 114 parts in 6 minutes. At this rate, how many parts can the machinist produce in 15 minutes?

12. RECIPES A recipe that makes 8 jumbo blueberry muffins calls for $1\frac{1}{2}$ teaspoons of baking powder. How much baking powder is needed to make 3 dozen jumbo muffins?

Estimate the unit price for each item. Justify your answers.

13. $299 for 4 tires

14. 3 yards of fabric for $13.47

UTILITIES For Exercises 15 and 16, use the table that shows the average monthly electricity and water usage.

Family Name	Family Size	Electricity (kilowatt-hours)	Water (gal)
Melendez	4	1,560	3,500
Barton	6	2,130	6,400
Stiles	2	1,490	2,500

15. Which family uses about twice the amount of electricity per person than the other two families? Explain your reasoning.

16. Which family uses the least amount of water per person? Explain your reasoning.

6-3 Study Guide and Intervention

Rate of Change and Slope

- A rate of change is a rate that describes how one quantity changes in relation to another.
- Slope tells how steep the line is.
- Slope is given by the formula $\dfrac{\text{change in } y}{\text{change in } x}$ or $\dfrac{\text{vertical change}}{\text{horizontal change}}$.

Example 1 Find the rate of change for the table.

Students	Number of Textbooks
5	15
10	30
15	45
20	60

The change in the number of textbooks is 15 while the change in the number of students is 5.

$\dfrac{\text{change in number of textbooks}}{\text{change in number of students}} = \dfrac{15 \text{ textbooks}}{5 \text{ students}}$ The number of textbooks increased by 15 for every 5 students.

$= \dfrac{3 \text{ textbooks}}{1 \text{ student}}$ Write as a unit rate.

So, the number of textbooks increases by 3 textbooks per student.

Example 2 The band boosters are selling T-shirts at a linear rate. By 8 P.M., they had sold 25 T-shirts. By 10 P.M., they had sold 45 T-shirts. Find the slope of the line. Explain what the slope represents.

$\dfrac{\text{change in number of T-shirts}}{\text{change in time}} = \dfrac{45 - 25}{10 - 8}$ Definition of slope.

$= \dfrac{20}{2}$ Simplify.

$= 10$

The slope is 10 and it means that the shirts are selling at a rate of 10 shirts per hour.

Exercises

Find the rate of change for each table.

1.

Side Length	Perimeter
1	4
2	8
3	12
4	16

2.

Time (in hours)	Distance (in miles)
2	120
4	240
6	360
8	480

3. The temperature at 10 A.M. was 72°F and at 2 P.M. was 88°F. Find the slope of the line. Explain what the slope represents.

6-3 **Practice**

Rate of Change and Slope

Find the rate of change for each table.

1.

Baby Age	Weight
0 months	0 pounds
3 months	12 pounds
6 months	24 pounds
9 months	36 pounds

2.

Number of Hours Worked	Money Earned ($)
4	80
6	120
8	160
10	200

3.

Days	Plant Height (in.)
7	4
14	11
21	18
28	25

4.

Months	Money Spent on Cable TV
2	82
4	164
6	246
8	328

Find the rate of change for each graph.

5. **Students in Mr. Muni's Class**

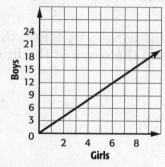

6. **Jewelry Making**

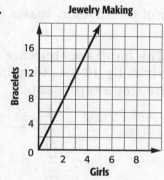

7. Graph the data. Then find the slope of the line. Explain what the slope represents.

Feet	1	2	3	4	5	6
Yards	3	6	9	12	15	18

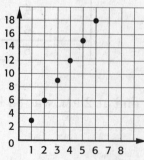

88

6-4 Study Guide and Intervention

Measurement: Changing Customary Units

Customary Units		
Length	Weight	Capacity
1 foot (ft) = 12 inches (in.)	1 pound (lb) = 16 ounces (oz)	1 cup (c) = 8 fluid ounces (fl oz)
1 yard (yd) = 3 feet	1 ton (T) = 2,000 pounds	1 pint (pt) = 2 cups
1 mile (mi) = 5,280 feet		1 quart (qt) = 2 pints
		1 gallon (gal) = 4 quarts

Example 1 $5\frac{1}{2}$ lb = ____?____ oz

To change from larger units to smaller units, multiply.

$5\frac{1}{2} \times 16 = 88$ Since 1 pound is 16 ounces, multiply by 16.

$5\frac{1}{2}$ pounds = 88 ounces

Example 2 28 fl oz = ____?____ c

To change from smaller units to larger units, divide.

$28 \div 8 = 3\frac{1}{2}$ Since 8 fluid ounces are in 1 cup, divide by 8.

28 fluid ounces = $3\frac{1}{2}$ cups

Exercises

Complete.

1. 5 lb = _____ oz

2. 48 in. = _____ ft

3. 6 yd = _____ ft

4. 7 qt = ____ pt

5. 8,000 lb = ____ T

6. $3\frac{1}{4}$ mi = _____ ft

7. 4 c = _____ fl oz

8. 6 c = _____ pt

9. $\frac{1}{2}$ gal = _____ qt

10. 3 ft = _____ in.

11. 9 qt = _____ gal

12. 30 fl oz = _____ c

13. 6,864 ft = ____ mi

14. 40 oz = _____ lb

15. 9 pt = _____ c

16. 18 ft = _____ yd

17. 11 pt = _____ qt

18. $2\frac{3}{4}$ T = _____ lb

Lesson 6-4

6-4 Practice

Measurement: Changing Customary Units

Complete.

1. 4 c = ___ fl oz

2. 5 c = ___ pt

3. 3 lb = ___ oz

4. 24 ft = ___ yd

5. $1\frac{1}{2}$ pt = ___ c

6. 64 oz = ___ lb

7. 4 mi = ___ ft

8. $2\frac{3}{4}$ mi = ___ ft

9. 3,000 lb = ___ T

10. 5 gal = ___ qt

11. $3\frac{1}{4}$ qt = ___ pt

12. $4\frac{5}{8}$ T = ___ lb

13. $3\frac{1}{2}$ gal = ___ qt

14. 7 c = ___ qt

15. 40 fl oz = ___ qt

16. 660 yd = ___ mi

17. 1.9 yd = ___ in.

18. $2\frac{1}{4}$ T = ___ oz

19. **SPORTS** The track surrounding a football field is $\frac{1}{4}$ mile long. How many yards long is the track?

20. **STRAWBERRIES** One quart of strawberries weighs about 2 pounds. About how many quarts of strawberries would weigh $\frac{1}{4}$ ton?

ANALYZE GRAPHS For Exercises 21–23, use the graph shown.

21. What does an ordered pair from this graph represent?

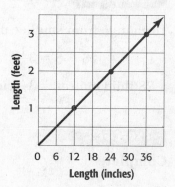

22. Write two sentences that describe the graph.

23. Use the graph to find the length in inches of a 1.5 foot iguana. Explain your reasoning.

6-5 Study Guide and Intervention

Measurement: Changing Metric Units

The table below is a summary of how to convert measures in the metric system.

	Larger Units → Smaller Units	Smaller Units → Larger Units
Units of Length (meter)	km to m – multiply by 1,000 m to cm – multiply by 100 m to mm – multiply by 1,000 cm to mm – multiply by 10	mm to cm – divide by 10 mm to m – divide by 1,000 cm to m – divide by 100 m to km – divide by 1,000
Units of Mass (kilogram)	kg to g – multiply by 1,000 g to mg – multiply by 1,000	mg to g – divide by 1,000 g to kg – divide by 1,000
Units of Capacity (liter)	kL to L – multiply by 1,000 L to mL – multiply by 1,000	mL to L – divide by 1,000 L to kL – divide by 1,000

Examples 1 Complete. 62 cm = ____ m

To convert from centimeters to meters, divide by 100.
62 ÷ 100 = 0.62
    62 cm = 0.62 m

Example 2 Complete. 2.6 kL = ____ L

To convert from kiloliters to liters, multiply by 1,000.
2.6 × 1,000 = 2,600
    2.6 kL = 2,600 L

Exercises

Complete.

1. 650 cm = ____ m

2. 57 kg = ____ g

3. 751 mg = ____ g

4. 8.2 L = ____ mL

5. 52 L = ____ kL

6. 892 mm = ____ m

7. 121.4 kL = ____ L

8. 0.72 cm = ____ mm

9. 67.3 g = ____ kg

10. 5.2 g = ____ mg

11. 0.05 m = ____ mm

12. 2,500 mg = ____ g

13. 32 mm = ____ cm

14. 96 m = ____ cm

Lesson 6-5

6-5 Practice

Measurement: Changing Metric Units

Complete.

1. 570 cm = _?_ m

2. 356 mm = _?_ m

3. 4.7 m = _?_ cm

4. 0.4 m = _?_ mm

5. 0.63 cm = _?_ mm

6. 0.18 mm = _?_ cm

7. 0.42 km = _?_ m

8. 0.09 km = _?_ mm

9. 0.13 km = _?_ cm

10. 27 kg = _?_ g

11. 8.3 g = _?_ mg

12. 257 mg = _?_ g

13. 486 g = _?_ kg

14. 55.5 g = _?_ kg

15. 68,700 mg = _?_ kg

16. 308 mL = _?_ L

17. 1.7 L = _?_ mL

18. 88 L = _?_ kL

19. 0.059 kL = _?_ L

20. 64,000 mL = _?_ L

21. 30,000 mL = _?_ kL

Order each set of measures from least to greatest.

22. 0.06 km, 47 m, 15,800 cm

23. 891 g, 7,800 mg, 0.5 kg

24. SPELUNKING The survey length of an underground cave is 0.914 kilometers. How many meters in length is this cave?

25. FOOD A 15-ounce box of granola contains 0.425 kilograms of cereal. How many grams of cereal are in the box of granola?

6-6 Study Guide and Intervention

Algebra: Solving Proportions

A **proportion** is an equation stating that two ratios are equivalent. Since rates are types of ratios, they can also form proportions. In a proportion, a **cross product** is the product of the numerator of one ratio and the denominator of the other ratio.

Example 1 Determine whether $\frac{2}{3}$ and $\frac{10}{15}$ form a proportion.

$\frac{2}{3} \overset{?}{=} \frac{10}{15}$ Write a proportion.

$2 \times 15 \overset{?}{=} 3 \times 10$ Find the cross products.

$30 = 30$ ✓   Multiply.

The cross products are equal, so the ratios form a proportion.

Example 2 Solve $\frac{8}{a} = \frac{10}{15}$.

$\frac{8}{a} = \frac{10}{15}$ Write the proportion.

$8 \times 15 = a \times 10$ Find the cross products.

$120 = 10a$ Multiply.

$\frac{120}{10} = \frac{10a}{10}$ Divide each side by 10.

$12 = a$ Simplify.

The solution is 12.

Exercises

Determine if the quantities in each pair of ratios are proportional. Explain.

1. $\frac{8}{10} = \frac{4}{5}$

2. $\frac{9}{4} = \frac{11}{6}$

3. $\frac{6}{14} = \frac{9}{21}$

4. $\frac{15}{12} = \frac{9}{6}$

5. $\frac{\$2.48}{4\ oz} = \frac{\$3.72}{6\ oz}$

6. $\frac{125\ mi}{5.7\ gal} = \frac{120\ mi}{5.6\ gal}$

Solve each proportion.

7. $\frac{y}{7} = \frac{16}{28}$

8. $\frac{5}{15} = \frac{15}{w}$

9. $\frac{20}{b} = \frac{70}{28}$

10. $\frac{52}{8} = \frac{m}{9}$

Lesson 6-6

6-6 Practice

Algebra: Solving Proportions

Determine if the quantities in each pair of ratios are proportional. Explain your reasoning.

1. 5 pounds of grass seed for 350 square feet and 8 pounds of grass seed for 560 square feet

2. 34 students from 8 schools and 25 students from 6 schools

Solve each proportion.

3. $\dfrac{5}{6} = \dfrac{a}{36}$

4. $\dfrac{k}{8} = \dfrac{8}{16}$

5. $\dfrac{7}{c} = \dfrac{14}{38}$

6. $\dfrac{4}{9} = \dfrac{40}{x}$

7. $\dfrac{12}{d} = \dfrac{5}{7}$

8. $\dfrac{6}{m} = \dfrac{42}{7}$

9. $\dfrac{n}{3.2} = \dfrac{3}{8}$

10. $\dfrac{2.8}{7.7} = \dfrac{z}{4.4}$

11. $\dfrac{1.5}{3.5} = \dfrac{4.5}{y}$

12. **CONDIMENTS** A store sells a 9-ounce jar of mustard for $1.53 and a 15-ounce jar for $2.55. Is the cost of the mustard proportional to the number of ounces for each jar? Explain your reasoning.

13. **SCIENCE** There are 113.2 grams in 4 ounces of compound. How many grams are in 5 ounces of compound?

14. **FURNITURE** A furniture company has 15 trucks that make about 120 deliveries each day. The company is expanding and expects an additional 40 deliveries each day. Write and solve a proportion to find how many more trucks are needed so the truck-to-delivery ratio remains the same.

15. **CHARITY** Karthik spent $35 of his allowance and gave $5 to a charity. If the number of dollars he spends is proportional to the number of dollars he gives to a charity, how much of a $100-allowance will he give to a charity?

6-7 Study Guide and Intervention

Problem-Solving Investigation: Draw a Diagram

When solving problems, draw a diagram to show what you have and what you need to find.

Example CARNIVAL Jim has to reach a target at a carnival game to win a prize. After 3 throws he has gone 75 feet, which is $\frac{3}{4}$ of the way to the target. How far away is the target?

Understand We know that 75 feet is $\frac{3}{4}$ of the way to the target.

Plan Draw a diagram to show the distance already thrown and the fraction it represents.

Solve

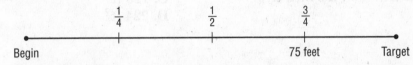

If $\frac{3}{4}$ of the distance is 75 feet, then $\frac{1}{4}$ of the distance is 25 feet. So, the missing $\frac{1}{4}$ must be another 25 feet.

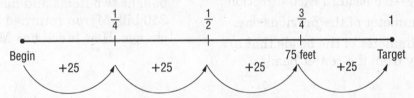

The total distance that Jim must throw to hit the target is 100 feet.

Check Since $\frac{3}{4}$ of the total distance is 75 feet, the equation $\frac{3}{4}x = 75$ represents this problem. Solving, we get $x = 100$ feet. So, the solution checks.

Exercises

1. **SALES** Sharon wants to buy a new car. She has saved up $ 1,500, which is approximately $\frac{1}{5}$ of the price of the car. How much does she need to save in order to buy the new car?

2. **TRAVEL** The Jones family has traveled 360 miles. They are $\frac{4}{5}$ of the way to their destination. How far away is their destination from where they started?

Lesson 6-7

6-7 Practice

Problem-Solving Investigation: Draw a Diagram

Mixed Problem Solving

Use the draw a diagram strategy to solve Exercises 1 and 2.

1. **ANTS** An ant went 2 meters away from its nest searching for food. The next time, the ant went 3 meters away. Each successive time the ant leaves the nest to search for food, the ant travels the sum of the two previous times. How far will the ant travel on his fifth trip?

2. **NECKLACES** The center bead of a pearl necklace has a 16 millimeter diameter. Each successive bead in each direction is $\frac{3}{4}$ the diameter of the previous one. Find the diameter of the beads that are three away from the center bead.

Use any strategy to solve Exercises 3–6. Some strategies are shown below.

PROBLEM-SOLVING STRATEGIES
• Work backward.
• Make an organized list.
• Eliminate possibilities.
• Draw a diagram.

3. **TALENT SHOW** At a talent show, 60% of the acts were singing. One-third of the remaining acts were instrumental. If 12 acts were instrumental, how many acts were in the talent show?

4. **GEOMETRY**
Miss Greenwell is adding 4 feet to the length and width of her rectangular garden as shown in the diagram. How much additional area will the garden have?
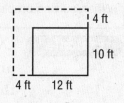
A. 16 ft^2
B. 104 ft^2
C. 120 ft^2
D. 224 ft^2

5. **YARD SALE** Myron has sold $18.50 worth of items at his yard sale. A neighbor bought two items and handed Myron a $10 bill. Myron returned $7.75 in change. How much has Myron now sold?

6. **COUNTRIES** The table shows the total land area of five countries.

Country	Total Area
Brazil	8.5 million sq km
Canada	10.0 million sq km
China	9.6 million sq km
Russia	17.1 million sq km
United States	9.6 million sq km

Estimate how much more total area Russia has than China.

Lesson 6-8

6-8 Study Guide and Intervention
Scale Drawings

A **scale drawing** represents something that is too large or too small to be drawn or built at actual size. Similarly, a **scale model** can be used to represent something that is too large or built too small for an actual-size model. The **scale** gives the relationship between the drawing/model measure and the actual measure.

Example On this map, each grid unit represents 50 yards. Find the distance from Patrick's Point to Agate Beach.

$$\begin{array}{ccc} & \textbf{Scale} & \begin{array}{c}\textbf{Patrick's Point}\\ \textbf{to Agate Beach}\end{array} \\ \text{map} \longrightarrow & \dfrac{1 \text{ unit}}{50 \text{ yards}} = & \dfrac{8 \text{ units}}{x \text{ yards}} \longleftarrow \text{map} \\ \text{actual} \longrightarrow & & \longleftarrow \text{actual} \end{array}$$

$$\begin{array}{lll} 1 \times x = & 50 \times 8 & \text{Cross products} \\ x = & 400 & \text{Simplify.} \end{array}$$

It is 400 yards from Patrick's Point to Agate Beach.

Exercises

Find the actual distance between each pair of cities. Round to the nearest tenth if necessary.

	Cities	Map Distance	Scale	Actual Distance
1.	Los Angeles and San Diego, California	6.35 cm	1 cm = 20 mi	
2.	Lexington and Louisville, Kentucky	15.6 cm	1 cm = 5 mi	
3.	Des Moines and Cedar Rapids, Iowa	16.27 cm	2 cm = 15 mi	
4.	Miami and Jacksonville, Florida	11.73 cm	$\frac{1}{2}$ cm = 20 mi	

Suppose you are making a scale drawing. Find the length of each object on the scale drawing with the given scale. Then find the scale factor.

5. an automobile 16 feet long; 1 inch:6 inches

6. a lake 85 feet across; 1 inch = 4 feet

7. a parking lot 200 meters wide; 1 centimeter:25 meters

8. a flag 5 feet wide; 2 inches = 1 foot

6-8 **Practice**

Scale Drawings

For Exercises 1–3, use the diagram of a section of the art museum shown. Use a ruler to measure.

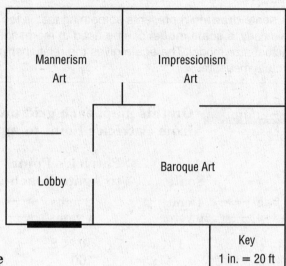

1. What is the actual length of the *Impressionism Art* room?

2. Find the actual dimensions of the *Baroque Art* room.

3. Find the scale factor for this blueprint.

Find the length of each model on the scale drawing with the given scale.

4. |← 8 ft →|

1 in. = 8 ft

5.

192 m

1 cm = 4 meters

6.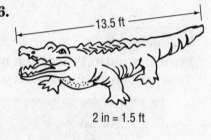

├── 13.5 ft ──┤

2 in = 1.5 ft

7. **SKYSCRAPER** A model of a skyscraper is made using a scale of 1 inch:75 feet. What is the height of the actual building if the height of the model is $19\frac{2}{5}$ inches?

8. **GEOGRAPHY** Salem and Eugene, Oregon, are 64 miles apart. If the distance on the map is $3\frac{1}{4}$ inches, find the scale of the map.

9. **PYRAMIDS** The length of a side of the Great Pyramid of Khufu at Giza, Egypt, is 751 feet. If you were to make a model of the pyramid to display on your desk, which would be an appropriate scale: 1 in. = 10 ft or 1 ft = 500 ft? Explain your reasoning.

6-9 Study Guide and Intervention

Fractions, Decimals, and Percents

Lesson 6-9

Example Write $4\frac{3}{8}\%$ as a fraction in simplest form.

$$4\frac{3}{8}\% = \frac{4\frac{3}{8}}{100} \qquad \text{Write a fraction.}$$

$$= 4\frac{3}{8} \div 100 \qquad \text{Divide.}$$

$$= \frac{35}{8} \div 100 \qquad \text{Write } 4\frac{3}{8} \text{ as an improper fraction.}$$

$$= \frac{35}{8} \times \frac{1}{100} \qquad \text{Multiply by the reciprocal of 100, which is } \frac{1}{100}.$$

$$= \frac{35}{800} \text{ or } \frac{7}{160} \qquad \text{Simplify.}$$

Example 2 Write $\frac{5}{16}$ as a percent.

$$\frac{5}{16} = \frac{n}{100} \qquad \text{Write a proportion using } \frac{n}{100}.$$

$$500 = 16n \qquad \text{Find the cross products.}$$

$$\frac{500}{16} = \frac{16n}{16} \qquad \text{Divide each side by 16.}$$

$$31\frac{1}{4} = n \qquad \text{Simplify.}$$

So, $\frac{5}{16} = 31\frac{1}{4}\%$ or 31.25%.

Exercises

Write each percent as a fraction in simplest form.

1. 60%

2. $68\frac{3}{4}\%$

3. $27\frac{1}{2}\%$

4. 37.5%

Write each fraction as a percent. Round to the nearest hundredth if necessary.

5. $\frac{2}{5}$

6. $\frac{5}{8}$

7. $\frac{9}{16}$

8. $\frac{2}{3}$

6-9 Practice

Fractions, Decimals, and Percents

Write each percent as a fraction in simplest form.

1. 37.5%   2. 5.8%   3. 43.75%   4. 52.5%

5. $83\frac{1}{3}\%$ 6. $66\frac{2}{3}\%$ 7. 135%   8. 0.01% $\frac{1}{10,000}$

Write each fraction as a percent. Round to the nearest hundredth if necessary.

9. $\frac{13}{20}$ 10. $\frac{9}{25}$ 11. $\frac{7}{8}$ 12. $\frac{39}{40}$

13. $\frac{5}{9}$ 14. $\frac{6}{7}$ 15. $\frac{2}{1}$ 16. $\frac{1}{1000}$

Replace each • with >, < or = to make a true statement.

17. $\frac{3}{16}$ • 24%   18. 0.775 • $\frac{31}{40}$ 19. 16% • 0.016

Order each set of numbers from least to greatest.

20. 0.6, 23%, 0.07, $\frac{2}{3}$ 21. $\frac{4}{5}\%$, 0.37, $\frac{1}{4}$, 0.4

22. **SAVINGS** Kayla has 14.5% of her salary placed into an Individual Retirement Account. What fraction is this?

23. **INTERNET** At home, 2 out of 5 people have access to broadband technology. What percent is this?

24. **SPORTS** A golfer made par on 13 of 18 holes. To the nearest tenth, on what percent of the holes did he make par?

ANALYZE TABLES For Exercises 25 and 26, use the table that shows the percent of households with the listed appliance.

25. What fraction of households have a clothes dryer?

26. Approximately 34 out of 67 households have a coffeemaker. Is this greater or less than the percent of households with a dishwasher? Explain.

Appliance	Percent of Households
Refrigerator	99.3%
Washing Machine	82.0%
Dryer	77.8%
Dishwasher	56.0%

7-1 Study Guide and Intervention

Percent of a Number

You can use a proportion or multiplication to find the percent of a number.

Example 1 Find 25% of 80.

$25\% = \dfrac{25}{100}$ or $\dfrac{1}{4}$ Write 25% as a fraction, and reduce to lowest terms.

$\dfrac{1}{4}$ of $80 = \dfrac{1}{4} \times 80$ or 20   Multiply.

So, 25% of 80 is 20.

Example 2 What number is 15% of 200?

15% of $200 = 15\% \times 200$ Write a multiplication expression.

$\qquad\qquad\quad = 0.15 \times 200$ Write 15% as a decimal.

$\qquad\qquad\quad = 30$ Multiply.

So, 15% of 200 is 30.

Exercises

Find each number.

1. Find 20% of 50.

2. What is 55% of $400?

3. 5% of 1,500 is what number?

4. Find 190% of 20.

5. What is 24% of $500?

6. 8% of $300 is how much?

7. What is 12.5% of 60?

8. Find 0.2% of 40.

9. Find 3% of $800.

10. What is 0.5% of 180?

11. 0.25% of 42 is what number?

12. What is 0.02% of 280?

Lesson 7-1

7-1 Practice

Percent of a Number

Find each number. Round to the nearest hundredth if necessary.

1. 55% of 140   **2.** 40% of 123   **3.** 37% of $150

4. 25% of 96   **5.** 11% of $333   **6.** 99% of 14

7. 140% of 30   **8.** 165% of 10   **9.** 150% of 150

10. 225% of 16   **11.** 106% of $40   **12.** 126% of 350

13. 4.1% of 30   **14.** 8.9% of 75   **15.** 24.2% of $120

16. 97.5% of 80

17. SALES Mr. Redding sells vehicles to 20% of the people that come to the sales lot. If 65 people came to the lot last month, how many vehicles did he sell?

Find each number. Round to the nearest hundredth if necessary.

18. $\frac{5}{6}$% of 600   **19.** $30\frac{1}{3}$% of 3   **20.** 1,000% of 87

21. 100% of 56   **22.** 0.25% of 150   **23.** 0.7% of 50

ANALYZE TABLES For Exercises 24–26, use the table that shows the percents of blood types of 145 donors during a recent blood drive.

Blood Type	Percent
O	45%
A	40%
B	11%
AB	4%

24. Write a proportion that can be used to find how many donors had type B blood. Then solve. Round to the nearest whole if necessary.

25. How many donors did *not* have type O blood? Round to the nearest whole if necessary.

26. Which blood type had less than 10 donors?

7-2 Study Guide and Intervention

The Percent Proportion

A percent proportion compares part of a quantity to a whole quantity for one ratio and lists the percent as a number over 100 for the other ratio.

$$\frac{part}{whole} = \frac{percent}{100}$$

Example 1 **What percent of 24 is 18?**

$\dfrac{part}{whole} = \dfrac{percent}{100}$ \qquad Percent proportion

Let $n\%$ represent the percent.

$\dfrac{18}{24} = \dfrac{n}{100}$ \qquad Write the proportion.

$18 \times 100 = 24 \times n$ \qquad Find the cross products.

$1{,}800 = 24n$ \qquad Simplify.

$\dfrac{1{,}800}{24} = \dfrac{24n}{24}$ \qquad Divide each side by 24.

$75 = n$

So, 18 is 75% of 24.

Example 2 **What number is 60% of 150?**

$\dfrac{part}{whole} = \dfrac{percent}{100}$ \qquad Percent proportion

Let a represent the part.

$\dfrac{a}{150} = \dfrac{60}{100}$ \qquad Write the proportion.

$a \times 100 = 150 \times 60$ \qquad Find the cross products.

$100a = 9{,}000$ \qquad Simplify.

$\dfrac{100a}{100} = \dfrac{9{,}000}{100}$ \qquad Divide each side by 100.

$a = 90$

So, 90 is 60% of 150.

Exercises

Find each number. Round to the nearest tenth if necessary.

1. What number is 25% of 20?

2. What percent of 50 is 20?

3. 30 is 75% of what number?

4. 40% of what number is 36?

5. What number is 20% of 625?

6. 12 is what percent of 30?

Lesson 7-2

7-2 Practice

The Percent Proportion

Find each number. Round to the nearest tenth if necessary.

1. What percent of 65 is 13?   2. $4 is what percent of $50?   3. What number is 35% of 22?

4. 14% of 81 is what number? 5. 13 is 26% of what number? 6. 55 is 40% of what number?

7. What percent of 45 is 72?   8. 1% of what number is 7?   9. 33 is 50% of what number?

10. What number is 3% of 100?   11. What percent of 200 is 0.5?

12. What number is 0.4% of 20?   13. What number is 6.1% of 60?

14. What percent of 34 is 34?   15. 10.4% of what number is 13?

16. **ALLOWANCE** Monica has $3 in her wallet. If this is 10% of her monthly allowance, what is her monthly allowance?

17. **WEDDING** Of the 125 guests invited to a wedding, 104 attended the wedding. What percent of the invited guests attended the wedding?

18. **CAMERA** The memory card on a digital camera can hold about 430 pictures. Melcher used 18% of the memory card while taking pictures at a family reunion. About how many pictures did Melcher take at the family reunion? Round to the nearest whole number.

OCEANS For Exercises 19 and 20, use the table shown.

19. The area of the Indian Ocean is what percent of the area of the Pacific Ocean? Round to the nearest whole percent.

Ocean	Area (square miles)
Pacific	64 million
Atlantic	32 million
Indian	25 million

Source: World Atlas

20. If the area of the Arctic Ocean is 16% of the area of the Atlantic Ocean, what is the area of the Arctic Ocean? Round to the nearest whole million.

7-4 Study Guide and Intervention

Algebra: The Percent Equation

> To solve any type of percent problem, you can use the **percent equation**, part = percent · base, where the percent is written as a decimal.

Example 1 **600 is what percent of 750?**

600 is the part and 750 is the whole. Let n represent the percent.

$$\underbrace{\text{part}}_{} = \underbrace{\text{percent}}_{} \cdot \underbrace{\text{whole}}_{}$$

$600 =$	n	$\cdot$ 750	Write an equation.
$\dfrac{600}{750} =$	$\dfrac{750n}{750}$		Divide each side by 750.
$0.8 = n$			Simplify.
$80\% = n$			Write 0.8 as a percent.

So, 600 is 80% of 750.

Example 2 **45 is 90% of what number?**

45 is the part and 90% or 0.9 is the percent. Let n represent the whole.

$$\underbrace{\text{part}}_{} = \underbrace{\text{percent}}_{} \cdot \underbrace{\text{whole}}_{}$$

$45 =$	0.9	$\cdot$ n	Write an equation.
$\dfrac{45}{0.9} =$	$\dfrac{0.9n}{0.9}$		Divide each side by 0.9.
$50 = n$			The whole is 50.

So, 45 is 90% of 50.

Exercises

Write an equation for each problem. Then solve. Round to the nearest tenth if necessary.

1. What percent of 56 is 14?

2. 36 is what percent of 40?

3. 80 is 40% of what number?

4. 65% of what number is 78?

5. What percent of 2,000 is 8?

6. What is 110% of 80?

7. 85 is what percent of 170?

8. Find 30% of 70.

Lesson 7-4

7-4) Practice

Algebra: The Percent Equation

Write an equation for each problem. Then solve. Round to the nearest tenth if necessary.

1. What number is 27% of 52?

2. Find 41% of 48.

3. What percent of 88 is 33?

4. 8 is what percent of 18?

5. What number is 33% of 360?

6. What percent of 62 is 58?

7. 55 is what percent of 100?

8. 22% of what number is 24.2?

9. 19 is 50% of what number?

10. 25 is 32% of what number?

11. 40% of what number is 28?

12. 30 is what percent of 60?

13. What percent of 5 is 2?

14. 44% of 10 is what number?

15. Find 110% of 88.

16. What number is 60% of 21.8?

17. What percent of 180 is 210?

18. 220 is 95.3% of what number?

19. **BASEBALL** A baseball player was at bat 473 times during the regular season. If he made a hit 31.5% of the times he was at bat, how many hits did he make during the regular season? Round to the nearest whole number if necessary.

ANALYZE GRAPHS For Exercises 20 and 21, use the graph shown. The total enrollment at Central High School is 798 students.

20. About what percent of the students at Central High are freshmen? Round to the nearest tenth if necessary.

21. About what percent of the students at Central High are seniors? Round to the nearest tenth if necessary.

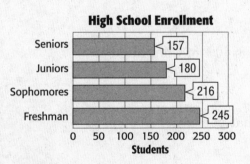

High School Enrollment

Seniors	157
Juniors	180
Sophomores	216
Freshman	245

0 50 100 150 200 250 300
Students

7-5 Study Guide and Intervention

Problem-Solving Investigation:
Determine Reasonable Answers

When solving problems, often times it is helpful to determine reasonable answers by using rounding and estimation. Checking answers with a calculator is always helpful in determining if the answer found is in fact reasonable.

Example SALES TAX There is 4.8% sales tax on all clothing items purchased. Danielle wants to buy a shirt, which costs $18.95. Danielle figures that if she has $20 she will have enough to buy the shirt. After adding in sales tax, is $20 a reasonable amount for Danielle to bring?

Understand	The cost of the shirt is $18.95. Sales tax is 4.8%. Danielle has $20.
Plan	Round $18.95 to $19.00 and 4.8% to 5%. Then use mental math to find 5% of $19.00.
Solve	Round $18.95 to $19.00
	Round 4.8% to 5%
	10% of $19.00 = 0.1 × 19 or $1.90 Use mental math. 10% = 0.1
	Round $1.90 to $2.00
	5% is $\frac{1}{2}$ of 10%
	So $\frac{1}{2}$ of $2.00 is $1.00 $1.00 is the amount of sales tax.
	$19.00 + $1.00 = $20.00 Add $1.00 to $19.00.
	So $20 is a reasonable amount of money for Danielle to bring to pay for the shirt.
Check	Use a calculator to check.
	0.048 × 18.95 = 0.9096
	Since 0.9096 is close to 1, the answer is reasonable.

Exercises

1. **TIP** The total bill at a restaurant for a family of 5 is $64.72. They want to leave a 20% tip. They decide to leave $10.00. Is this estimate reasonable? Explain your reasoning.

2. **TELEVISION** A recent survey shows that 67% of students watch 3 or more hours of television a night. Suppose there are 892 students in your school. What would be a reasonable estimate of the number of students in your school who watch 3 or more hours of television a night? Explain your reasoning.

7-5 Practice

Problem-Solving Investigation:
Determine Reasonable Answers

Mixed Problem Solving

For Exercises 1 and 2, determine a reasonable answer.

1. **HOMES** In a retirement village, 86% of the residents own their home. If the village has 540 homes, how many homes are owned by the residents, about 250, 350, or 450?

2. **ANALYZE GRAPHS** The graph shows how the Forenzo family spent their money on their summer vacation. Is 25% a reasonable estimate of how much money they spent on dining? Justify your answer.

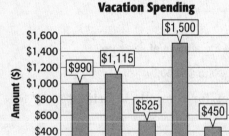

Vacation Spending

Use any strategy to solve Exercises 3–6. Some strategies are shown below.

PROBLEM-SOLVING STRATEGIES
• Guess and check.
• Make an organized list.
• Determine reasonable answers.

3. **NUMBER SENSE** 12 is added to 25% of a number. The result is 30. What is the number?

4. **ANALYZE GRAPHS** The graph shows the percent of community attendance during a little league season. Is 90% a reasonable estimate for the percent of community attendance for September? Explain.

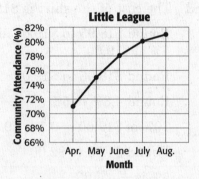

5. **TRAVEL** Cecil averages 31 miles per gallon when driving his car on the highway to visit friends 461 miles away. If he filled the 16-gallon gasoline tank before leaving and did not buy any gasoline along the way, about how many gallons of gasoline are left in the tank when he arrives?

6. **FABRIC** Mrs. Tillman is making identical dresses for her three granddaughters. She needs $2\frac{1}{8}$ yards of fabric for each dress. If she purchased $8\frac{1}{2}$ yards of fabric, how much fabric will be leftover?

7-6 Study Guide and Intervention
Percent of Change

Lesson 7-6

A **percent of change** is a ratio that compares the change in quantity to the original amount. If the original quantity is increased, it is a **percent of increase**. If the original quantity is decreased, it is a **percent of decrease**.

Example 1 Last year, 2,376 people attended the rodeo. This year, attendance was 2,950. What was the percent of change in attendance to the nearest whole percent?

Since this year's attendance is greater than last year's attendance, this is a percent of increase.

The amount of increase is 2,950 – 2,376 or 574.

percent of increase $= \dfrac{\text{amount of increase}}{\text{original amount}}$ ← new amount – original amount

$= \dfrac{574}{2,376}$ Substitution

≈ 0.24 or 24% Simplify.

Rodeo attendance increased by about 24%.

Example 2 John's grade on the first math exam was 94. His grade on the second math exam was 86. What was the percent of change in John's grade to the nearest whole percent?

Since the second grade is less than the first grade, this is a percent of decrease. The amount of decrease is 94 − 86 or 8.

percent of decrease $= \dfrac{\text{amount of decrease}}{\text{original amount}}$ ← original amount – new amount

$= \dfrac{8}{94}$ Substitution

≈ 0.09 or 9% Simplify.

John's math grade decreased by about 9%.

Exercises

Find each percent of change. Round to the nearest whole percent if necessary. State whether the percent of change is an *increase* or *decrease*.

1. original: 4
   new: 5

2. original: 1.0
   new: 1.3

3. original: 15
   new: 12

4. original: $30
   new: $18

5. original: 60
   new: 63

6. original: 160
   new: 136

7. original: 7.7
   new: 10.5

8. original: 9.6
   new: 5.9

7-6 Practice

Percent of Change

Find each percent of change. Round to the nearest whole percent if necessary. State whether the percent of change is an *increase* or *decrease*.

1. 8 feet to 10 feet

2. 136 days to 85 days

3. $0.32 to $0.37

4. 62 trees to 31 trees

5. 51 meters to 68 meters

6. 16.5 grams to 24.8 grams

7. 0.55 minutes to 0.1 minutes

8. $180 to $210

9. 2.9 months to 4.9 months

10. $\frac{1}{4}$ to $\frac{3}{8}$

11. $\frac{1}{6}$ to $\frac{1}{3}$

12. $\frac{4}{3}$ to $\frac{1}{3}$

13. SURGERY Recent developments in surgical procedures change the average healing time for some operations from 8 weeks to 3 weeks.

14. ROADS The city added an extra lane in each direction to the 5-lane road.

GEOMETRY For Exercises 15 and 16, refer to the rectangle shown. Suppose the width is decreased by 3 inches.

15. Find the percent change in the perimeter.

16. Find the percent change in the area.

$\ell = 4$ in.

$w = 6$ in.

ANALYZE TABLES For Exercises 17 and 18, refer to the table that shows the average monthly rainfall during the first six months of the year for Singapore.

17. Between which two consecutive months is the percent of decrease the greatest? What is the percent change? Round to the nearest whole percent.

18. Between which two consecutive months is the percent of increase the least? What is the percent change? Round to the nearest whole percent.

Month	Average Rainfall (inches/month)
January	9.4
February	6.5
March	6.8
April	6.6
May	6.7
June	6.4

7-7 Study Guide and Intervention

Sales Tax and Discount

> **Sales tax** is a percent of the purchase price and is an amount paid in addition to the purchase price.
> **Discount** is the amount by which the regular price of an item is reduced.

Example 1 SOCCER **Find the total price of a $17.75 soccer ball if the sales tax is 6%.**

Method 1

First, find the sales tax.
6% of $17.75 = 0.06 · 17.75
≈ 1.07
The sales tax is $1.07.

Next, add the sales tax to the regular price.

1.07 + 17.75 = 18.82

The total cost of the soccer ball is $18.82.

Method 2

100% + 6 % = 106%   Add the percent of tax
to 100%.
The total cost is 106% of the regular price.

106% of $17.75 = 1.06 · 17.75
≈ 18.82

Example 2 TENNIS **Find the price of a $69.50 tennis racket that is on sale for 20% off.**

First, find the amount of the discount d.

$\underbrace{\text{part}} = \underbrace{\text{percent}} \cdot \underbrace{\text{whole}}$

$d = 0.2 \cdot 69.50$ Use the percent equation.
$d = 13.90$ The discount is $13.90.

So, the sale price of the tennis racket is $69.50 − $13.90 or $55.60.

Exercises

Find the total cost or sale price to the nearest cent.

1. $22.95 shirt; 7% sales tax

2. $39.00 jeans; 25% discount

3. $35 belt; 40% discount

4. $115.48 watch; 6% sales tax

5. $16.99 book; 5% off

6. $349 television; 6.5% sales tax

Lesson 7-7

7-7 Practice

Sales Tax and Discount

Find the total cost or sale price to the nearest cent.

1. $18 haircut; 10% discount   2. $299 lawn mower; 5% tax   3. $9.99 meal; 25% discount

4. $149 guitar; 20% discount   5. $15.75 music CD; 4% tax

6. $24 gym bag; 8% tax   7. $32.88 jacket; 50% discount

8. $3.45 coffee; 33% discount   9. $9.99 chair; $8\frac{1}{2}$% tax

Find the original price to the nearest cent.

10. bracelet: discount, 40%   11. bicycle: discount, 35%
      sale price, $13.80   sale price, $79

12. **TICKETS** State residents get discounts at various theme parks throughout the state.
One theme park charges a state resident $51.70. If this price represents a 15% discount
from the regular adult admission, find the cost of a regular adult admission to the
nearest cent.

13. **TRUCKS** What is the sales tax on a $17,500 truck if the tax rate is 6%?

COMPUTERS For Exercises 14–16, use the following information.

Lionel is buying a computer that normally sells for $890. The state sales tax is 6%.

14. What is the total cost of the computer including tax?

15. If the computer is on sale with a 10% discount, what is the sale price of the computer
before adding the sales tax?

16. What is the sales tax on the discounted price?

7-8 Study Guide and Intervention

Simple Interest

Simple interest is the amount of money paid or earned for the use of money. To find simple interest I, use the formula $I = prt$. Principal p is the amount of money deposited or invested. Rate r is the annual interest rate written as a decimal. Time t is the amount of time the money is invested in years.

Example 1 Find the simple interest earned in a savings account where $136 is deposited for 2 years if the interest rate is 7.5% per year.

$I = prt$ Formula for simple interest

$I = 136 \cdot 0.075 \cdot 2$ Replace p with $136, r with 0.075, and t with 2.

$I = 20.40$ Simplify.

The simple interest earned is $20.40.

Example 2 Find the simple interest for $600 invested at 8.5% for 6 months.

6 months $= \dfrac{6}{12}$ or 0.5 year   Write the time as years.

$I = prt$ Formula for simple interest

$I = 600 \cdot 0.085 \cdot 0.5$ $p = $600, $r = 0.085, $t = 0.5$

$I = 25.50$ Simplify.

The simple interest is $25.50.

Exercises

Find the interest earned to the nearest cent for each principal, interest rate, and time.

1. $300, 5%, 2 years

2. $650, 8%, 3 years

3. $575, 4.5%, 4 years

4. $735, 7%, $2\frac{1}{2}$ years

5. $1,665, 6.75%, 3 years

6. $2,105, 11%, $1\frac{3}{4}$ years

7. $903, 8.75%, 18 months

8. $4,275, 19%, 3 months

Lesson 7-8

7-8 Practice

Simple Interest

Find the simple interest earned to the nearest cent for each principal, interest rate, and time.

1. $750, 7%, 3 years

2. $1,200, 3.5%, 2 years

3. $450, 5%, 4 months

4. $1,000, 2%, 9 months

5. $530, 6%, 1 year

6. $600, 8%, 1 month

Find the simple interest paid to the nearest cent for each loan, interest rate, and time.

7. $668, 5%, 2 years

8. $720, 4.25%, 3 months

9. $2,500, 6.9%, 6 months

10. $500, 12%, 18 months

11. $300, 9%, 3 years

12. $2,000, 20%, 1 year

13. **ELECTRONICS** Rita charged $126 for a DVD player at an interest rate of 15.9%. How much will Rita have to pay after 2 months if she makes no payments?

14. **VACATION** The average cost for a vacation is $1,050. If a family borrows money for the vacation at an interest rate of 11.9% for 6 months, what is the total cost of the vacation including the interest on the loan?

For Exercises 15–17, use the following information.

Robin has $2,500 to invest in a CD (certificate of deposit).

15. If Robin invests the $2,500 in the CD that yields 4% interest, what will the CD be worth after 2 years?

16. Robin would like to have $3,000 altogether. If the interest rate is 5%, in how many years will she have $3,000?

17. Suppose Robin invests the $2,500 for 3 years and earns $255. What was the rate of interest?

8-1 Study Guide and Intervention

Line Plots

A **line plot** is a diagram that shows the frequency of data on a number line.

Example 1 SHOE SIZE The table shows the shoe size of students in Mr. Kowa's classroom. Make a line plot of the data.

Shoe Sizes			
10	6	4	6
5	11	10	10
6	9	6	8
7	11	7	14
5	10	6	10

Step 1 Draw a number line. Because the smallest size is 4 and the largest size is 14, you can use a scale of 4 to 14 and an interval of 2.

Step 2 Put an "×" above the number that represents the shoe size of each student.

```
        ×           ×
        ×           ×
        ×           ×
      × × ×       × × ×
    × × × × × × × × × ×       ×
    ┼─┼─┼─┼─┼─┼─┼─┼─┼─┼─┼
    4   6   8   10  12  14
```

Example 2 Use the line plot in Example 1. Identify any clusters, gaps, or outliers and analyze the data by using these values. What is the range of data?

Many of the data cluster around 6 and 10. You could say that most of the shoe sizes are 6 or 10. There is a gap between 11 and 14, so there are no shoe sizes in this range. The number 14 appears removed from the rest of the data, so it would be considered an outlier. This means that the shoe size of 14 is very large and is not representative of the whole data set.

The greatest shoe size is 14, and the smallest is 4. The range is 14 – 4 or 10.

Exercises

PETS For Exercises 1–3 use the table at the right that shows the number of pets owned by different families.

Number of Pets			
2	1	2	0
3	1	1	2
8	3	1	4

1. Make a line plot of the data.

2. Identify any clusters, gaps, or outliers.

3. What is the range of the data?

Lesson 8-1

8-1 Practice

Line Plots

Display each set of data in a line plot.

1.

Weights of Dogs (pounds)				
21	12	33	14	17
8	30	18	15	25
14	21	14	19	12

2.

Quiz Scores				
88	94	83	94	90
99	78	88	94	84
90	88	96	86	93

3.

Miles Driven					
132	115	95	111	108	94
124	113	125	95	110	115
122	107	99	115	121	133

4.

Drying Time (minutes)					
15	16	13	14	15	16
14	16	13	16	15	14
14	13	16	15	14	15

RAINFALL For Exercises 5–9, analyze the line plot that shows the amount of daily rainfall in inches during 30 consecutive days in a rainy season.

Daily Rainfall (inches)

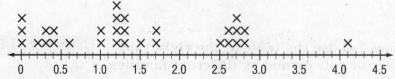

5. Find the range of the data.

6. How many days did it rain more than 1.0 inch?

7. What rainfall amount occurred most often?

8. Identify any clusters, gaps, or outliers.

8-2 Study Guide and Intervention

Measures of Central Tendency and Range

The **mean** is the sum of the data divided by the number of data items. The **median** is the middle number of the ordered data, or the mean of the middle two numbers. The **mode** is the number (or numbers) that occur most often. The mean, median, and mode are each **measures of central tendency**.

Example The table shows the number of hours students spent practicing for a music recital. Find the mean, median, and mode of the data.

Numbers of Hours Spent Practicing				
3	12	10	8	7
18	11	12	10	3
8	6	0	1	5
8	2	15	9	12

$$\text{mean} = \frac{3 + 12 + 10 + \ldots + 12}{20} = \frac{160}{20} \text{ or } 8.$$

To find the median, the data must be ordered.

0, 1, 2, 3, 3, 5, 6, 7, 8, 8, 8, 9, 10, 10, 11, 12, 12, 12, 15, 18

$$\frac{8 + 8}{2} = 8$$

To find the mode, look for the number that occurs most often. Since 8 and 12 each occur 3 times, the modes are 8 and 12.

Exercises

Find the mean, median, and mode for each set of data. Round to the nearest tenth if necessary.

1. 27, 56, 34, 19, 41, 56, 27, 25, 34, 56

2. 7, 3, 12, 4, 6, 3, 4, 8, 7, 3, 20

3. 1, 23, 4, 6, 7, 20, 7, 5, 3, 4, 6, 7, 11, 6

4. 3, 3, 3, 3, 3, 3, 3

5. 2, 4, 1, 3, 5, 6, 1, 1, 3, 4, 3, 1

6. 4, 0, 12, 10, 0, 5, 7, 16, 12, 10, 12, 12

Lesson 8-2

8-2 Practice

Measures of Central Tendency and Range

Find the mean, median, and mode for each set of data. Round to the nearest tenth if necessary.

1. Number of parking spaces used: 45, 39, 41, 45, 44, 64, 51

2. Prices of plants: $10, $8, $20, $25, $14, $39, $10, $10, $8, $16

3. Points scored during football season: 14, 20, 3, 9, 18, 35, 21, 24, 31, 12, 7

4. Golf scores: −3, −2, +1, +1, −1, −1, +2, −5

5. Percent increase: 3.3, 4.1, 3.9, 5.0, 3.5, 2.9, 3.9

6.

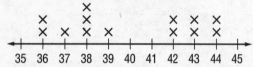

Dollars Spent Shopping

7. **CHILDREN** The table shows the number of children living at home in a neighborhood of 24 homes. Which measure best describes the data: mean, median, or mode? Explain.

Children at Home							
2	1	3	0	4	4	1	2
0	6	2	2	5	0	2	3
3	1	1	4	2	0	1	4

8. **WORK** The table shows the hours Sam worked each week during the summer. How many hours did he work during the twelfth week to average 20 hours per week?

Hours Worked					
18	24	20	19	15	21
20	19	18	22	22	?

8-3

Study Guide and Intervention

Stem-and-Leaf Plots

In a **stem-and-leaf plot**, the data are organized from least to greatest. The digits of the least place value usually form the **leaves**, and the next place value digits form the **stems**.

Example Make a stem-and-leaf plot of the data below. Then find the range, median, and mode of the data.
42, 45, 37, 46, 35, 49, 47, 35, 45, 63, 45

Order the data from least to greatest.

35, 35, 37, 42, 45, 45, 45, 46, 47, 49, 63

The least value is 35, and the greatest value is 63.
So, the tens digits form the stems, and the ones
digits form the leaves.

Stem	Leaf
3	5 5 7
4	2 5 5 5 6 7 9
5	
6	3

$6 \mid 3 = 63$

range: greatest value − least value = 63 − 35 or 28
median: middle value, or 45
mode: most frequent value, or 45

Exercises

Make a stem-and-leaf plot for each set of data. Then find the range, median, and mode of the data.

1. 15, 25, 16, 28, 1, 27, 16, 19, 28

2. 1, 2, 3, 2, 3, 1, 4, 2, 5, 7, 12, 11, 11, 3, 10

3. 3, 5, 1, 17, 11, 45, 17

4. 4, 7, 10, 5, 8, 12, 7, 6

8-3 Practice

Stem-and-Leaf Plots

Display each set of data in a stem-and-leaf plot.

1.

Test Scores				
78	99	83	92	90
94	88	88	94	87
70	86	85	86	93

2.

Weight of Male Lions (pounds)			
440	425	452	433
445	436	440	475
426	444	455	485
437	450	466	470

GOLD MEDALS For Exercises 3–5, use the stem-and-leaf plot that shows the number of gold medals won by each of the top 15 countries at the 2004 Summer Olympics.

Stem	Leaf
0	6 8 8 9 9 9 9
1	0 1 4 6 7
2	7
3	2 5 1|4 = 14 gold medals

3. Find the range of gold medals won.

4. Find the median and the mode of the data.

5. Based on the data, write one inference that can be made about the data.

PRESIDENTS For Exercises 6–10, use the stem-and-leaf plot that shows the age of each United States President at inauguration.

	Ages of U.S. Presidents at Inauguration
Stem	Leaf
4	2 3 6 6 7 8 9 9
5	0 0 1 1 1 1 2 2 4 4 4 4 5 5 5 5 6 6 6 7 7 7 7 7 8
6	0 1 1 1 2 4 4 5 8 9 4|1 = 41 years

6. How many presidents were under the age of 45 when inaugurated?

7. Find the ages of the youngest and oldest president at inauguration.

8. Find the range of the data.

9. Find the median and the mode of the data

10. Based on the data, in what age group were the majority of the presidents when inaugurated?

8-4 Study Guide and Intervention

Bar Graphs and Histograms

A **bar graph** is one method of comparing data by using solid bars to represent quantities. A **histogram** is a special kind of bar graph. It uses bars to represent the frequency of numerical data that have been organized into intervals.

Example 1 SIBLINGS **Make a bar graph to display the data in the table below.**

Student	Number of Siblings
Sue	1
Isfu	6
Margarita	3
Akira	2

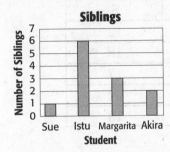

Step 1 Draw a horizontal and a vertical axis. Label the axes as shown. Add a title.

Step 2 Draw a bar to represent each student. In this case, a bar is used to represent the number of siblings for each student.

Example 2 SIBLINGS **The number of siblings of 17 students have been organized into a table. Make a histogram of the data.**

Number of Siblings	Frequency
0–1	4
2–3	10
4–5	2
6–7	1

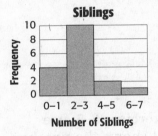

Step 1 Draw and label horizontal and vertical axes. Add a title.

Step 2 Draw a bar to represent the frequency of each interval.

Exercises

1. Make a bar graph for the data in the table.

Student	Number of Free Throws
Luis	6
Laura	10
Opal	4
Gad	14

2. Make a histogram for the data in the table.

Number of Free Throws	Frequency
0–1	1
2–3	5
4–5	10
6–7	4

8-4 Practice

Bar Graphs and Histograms

Select the appropriate graph to display each set of data: bar graph or histogram. Then display the data in the appropriate graph.

1.

Ages of Children Taking Swimming Lessons	
Age	Children
0–2	8
3–5	12
6–8	18
9–11	17
12–14	12
15–17	13

2.

Home Run Derby 2007 Round 1 Home Runs	
Player	Home Runs
Vladimir Guerrero	5
Alex Rios	5
Matt Holliday	5
Albert Pujols	4
Justin Morneau	4

Source: Baseball Almanac

POPULATION For Exercises 3–5, use the bar graph that shows the number of males and females in the world for the years 1970, 1980, 1990, 2000, 2005.

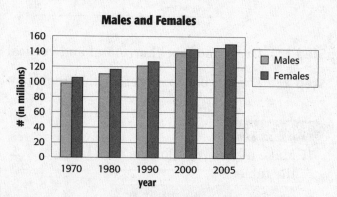

3. By how much did the number of females increase from 1970 to 1980?

4. By how much did the number of females increase from 2000 to 2005?

5. Between which years did the number of females increase the most?

8-5 Study Guide and Intervention

Problem-Solving Investigation: Use a Graph

When solving problems, a **graph** can show a visual representation of the situation and help you make conclusions about the particular set of data.

Example **POPULATION** The table shows the enrollment of Mill High School students over five years. Estimate the enrollment was for the 2010–2011 school year.

Mill High School Enrollment				
'05–'06	'06–'07	'07–'08	'08–'09	'09–'10
115	134	168	160	185

Understand You know the enrollment of students for five years. You need to estimate the enrollment for the 2010–2011 school year.

Plan Organize the data in a graph so that you can see a trend in the enrollment levels.

Solve

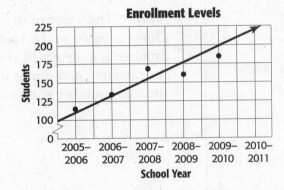

The graph shows that the enrollment increases over the years. By using the graph you can conclude that Mill High School had about 225 students enrolled for the 2010–2011 school year.

Check Draw a line through as close to as many points as possible. The estimate is close to the line so the answer is reasonable.

Exercises

1. **TEMPERATURE** The chart to the right shows the average December temperatures in Fahrenheit over four years. Predict the average temperature for the next year.

December Temperatures (F°)			
2002	2003	2004	2005
22°	17°	18°	16°

2. **POPULATION** Every five years the population of your neighborhood is recorded. What do you predict the population will be in 2010?

Neighborhood Population		
1995	2000	2005
2,072	2,250	2,376

Lesson 8-5

8-5 Practice

Problem-Solving Investigation: Use a Graph

Mixed Problem Solving

PITCHING For Exercises 1 and 2, use the graph that shows the amount of pitching practice time for Adam and Jordan during a particular week.

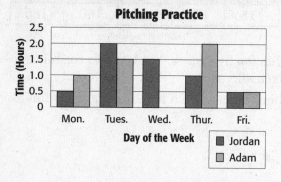

Pitching Practice

1. Who practiced more during the week and by how much time?

2. What was Adam's average practice time per day for the five days?

Use any strategy to solve Exercises 3–6. Some strategies are shown below.

PROBLEM-SOLVING STRATEGIES
• Guess and check.
• Look for a pattern.
• Make a graph.

3. **LAWN TOOLS** The bar graph shows the number of shovels and rakes sold during particular months at a hardware store. During which month was the number of rakes sold about twice the number of shovels sold?

Shovel and Rake Sales

4. **NUMBER THEORY** 42 is subtracted from 42% of a number. The result is 42. What is the number?

5. **MONEY** The value of the number of dimes is equal to the value of the number of quarters. If the total value of the quarters and dimes is $6.00, find the total number of coins.

6. **SKIING** Mrs. Roget is taking her family of 2 adults and 4 children skiing for the day. They need to rent ski equipment. What will it cost to ski for the day including equipment rental and lift tickets?

Daily Ski Costs		
Item	Adults	Children
Lift Ticket	$10.00	$8.00
Skis	$7.00	$4.25
Boots	$6.25	$4.25
Poles	$2.25	$1.75

8-6 Study Guide and Intervention

Using Graphs to Predict

A **line graph** shows trends over time and can be useful for predicting future events. A **scatter plot** displays two sets of data on a graph and can be useful for predictions by showing trends in the data.

Example Use the line graph of the Moralez family car trip shown below to answer the following questions.

1. After 250 miles, how much gas did the Moralez family have left?

> Draw a dotted line up from 250 m until it reaches the graph and then find the corresponding gas measure.

They will have about 5.5 g left.

2. How far can the Moralez family travel before they run out of gas?

> When they run out of gas, the tank will be at 0 so find where the line reaches 0.

They can travel about 410 miles.

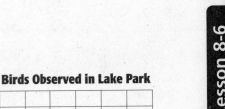

Gas Usage

Exercises

Use the scatter plot to answer the questions.

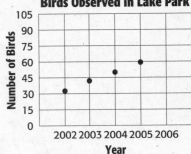

Birds Observed in Lake Park

1. How many birds were there in 2004?

2. What relationship do you see between the number of birds and year?

3. Predict the number of birds there were in the year 2001?

4. Predict the number of birds there will be in in the year 2006?

5. In what year do you think the bird population will reach 100?

Lesson 8-6

8-6 Practice

Using Graphs to Predict

WATER LEVEL For Exercises 1 and 2, use the graph that shows the level of rising water of a lake after several days of rainy weather.

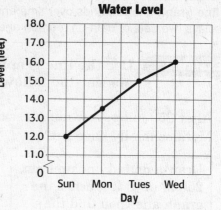

1. If the water continues to rise, predict the day when the water level will be above flood stage of 20.5 feet.

2. How many days did it take for the water level to rise 4 feet?

PROPERTY For Exercises 3–5, use the table that shows the property value per acre for five years.

3. Make a scatter plot of the data. Use the time on the horizontal axis and the property value on the vertical axis.

Property Value (per acre)	
Time	Value
2005	$14,000
2006	$16,600
2007	$18,900
2008	$21,500
2009	$24,000

4. Describe the relationship, if any, between the two sets of data.

5. Predict the property value per acre in 2010.

8-7 Study Guide and Intervention

Using Data to Predict

Data gathered by surveying a random sample of the population may be used to make predictions about the entire population.

Example 1 In a survey, 200 people from a town were asked if they thought the town needed more bicycle paths. The results are shown in the table. Predict how many of the 28,000 people in the town think more bicycle paths are needed.

More Bicycle Paths Needed?	
Response	**Percent**
yes	39%
no	42%
undecided	19%

Use the percent proportion.

$$\frac{part}{whole} = \frac{percent}{100}$$ Percent proportion

part of the population $\longrightarrow$

$$\frac{n}{28,000} = \frac{39}{100}$$ Let n represent the number.
Survey results: $39\% = \frac{39}{100}$

Whole population

$$100n = 38,000(39)$$ Cross products

$$n = 10,920$$ Simplify.

So, about 10,920 people in the town think more bicycle paths are needed.

Exercises

1. **VOTES** In a survey of voters in Binghamton, 55% of those surveyed said they would vote for Armas for city council. If 24,000 people vote in the election, about how many will vote for Armas?

2. **LUNCH** A survey shows that 43% of high school and middle school students buy school lunches. If a school district has 2,900 high school and middle school students, about how many buy school lunches?

3. **CLASS TRIP** Students of a seventh grade class were surveyed to find out how much they would be willing to pay to go on a class trip. 24% of the students surveyed said they would pay $21 to $30. If there are 360 students in the seventh grade class, about how many would be willing to pay for a trip that cost $21 to $30?

Lesson 8-7

8-7 Practice

Using Data to Predict

Match each situation with the appropriate equation or proportion.

1. 85% of commuters use the expressway. Predict how many commuters out of 750 commuters will use the expressway.

2. 750% of 85 is what number?

3. 85 commuters is what percent of 750 commuters?

a. $n = 0.85 \cdot 750$

b. $\dfrac{85}{750} = \dfrac{n}{100}$

c. $7.5 \cdot 85 = n$

4. **ESKIMOS** In the year 2000, the population of Alaska was about 627 thousand. Predict the number of Eskimos in Alaska if the Eskimo population was about 7.5% of the population of Alaska. Round to the nearest thousand.

5. **DOGS** A survey showed that about 40% of American households own at least one dog. Based on that survey, how many households in a community of 800 households own at least one dog?

CAR REPAIRS For Exercises 6–8, use the graph that shows the percent of all repairs for 3 car repair problems at a car repair shop.

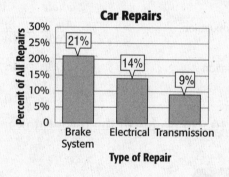

Car Repairs

6. Suppose a mechanic repairs 478 cars. Predict how many repairs will be made on transmissions.

7. For every 100 repairs, predict how many more repairs will be made on a brake system problem than on an electrical problem.

8. Predict the percent of repairs that will be one of the three problems in the graph.

130

8-8 Study Guide and Intervention

Using Sampling to Predict

In an **unbiased sample** the whole population is represented. In a **biased sample** one or more parts of the population are favored over the others.

Example 1 Look at the following table to determine the favorite sport of middle school students.

Favorite Sports of Middle School Students			
Basketball	Baseball	Football	Soccer
10	5	17	52

Based on the table, it would appear that soccer is the favorite sport of middle school students. However, suppose the data collected for this survey was taken at a World Cup soccer match. It can then be concluded that our sample is **biased** because students who are at a soccer match may be more likely to choose soccer as their favorite sport.

To receive an **unbiased** sample of middle school students, the sports survey could be completed at randomly selected middle schools throughout the country.

Exercises Determine whether the given situations represent a *biased* or *unbiased* sample. Then tell the type of sample.

1. Writers of a popular teen magazine want to write a story about which movies their readers like. The writers decide to interview the first 50 people that walk out of a movie theater.

2. The student council wanted to raise money for their school by selling homemade cookies during lunch time. To find out the favorite kind of cookie for the majority of their school, they conducted a survey. They gave the survey to 20 randomly selected students from each grade level.

3. To determine the most frequently used gas station, a researcher randomly selected every 10th person from a drive-through fast food restaurant and asked them where they last filled up with gas.

Lesson 8-8

8-8 Practice

Using Sampling to Predict

Determine if the sample method is valid (unbiased) and if so, use the results to make predictions. If the sample is not valid (biased), write *not valid* on the line and explain why.

1. A representative from the cable company randomly calls 100 households to determine the number of customers who receive movie channels. Of these, 15% do have movie channel access. If there are 2,300 customers total, how many can be expected to have the movie channels?

2. An electronics store just received a huge shipment of video games. Kenny has been put in charge of making sure the goods are not damaged. There are 350 boxes and 50 games in each box. Kenny decides to take the nearest 5 boxes and check for damages. He finds only 2 damaged games, so what can he predict for the total number of damaged games in the boxes?

3. Taylor was given the following problem:

   A researcher, who was trying to link after-school students from 20 different schools around the country, surveyed 50 children from each school. He found that 74% of students were involved in after-school sports. How many students surveyed were involved in sports?

   This is how Taylor solved the problem:

50	1000	It's valid because it
$\times\ 20$	$\times\ \ 74$	is a simple random
1,000	74,000	sample and there were
		74,000 students.

   Explain what Taylor did wrong.

8-9 Study Guide and Intervention

Misleading Statistics

Graphs can be misleading for many reasons: there is no title, the scale does not include 0; there are no labels on either axis; the intervals on a scale are not equal; or the size of the graphics misrepresents the data.

Example WEEKLY CHORES The line graphs below show the total hours Salomon spent doing his chores one month. Which graph would be best to use to convince his parents he deserves a raise in his allowance? Explain.

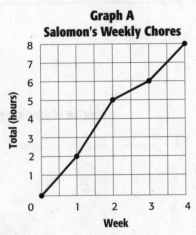

Graph A
Salomon's Weekly Chores

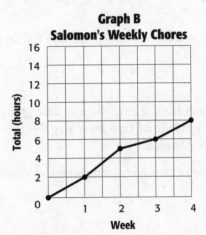

Graph B
Salomon's Weekly Chores

He should use graph A because it makes the total hours seem much larger.

Exercises

PROFITS For Exercises 1 and 2, use the graphs below. It shows a company's profits over a four-month period.

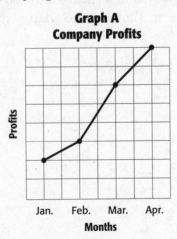

Graph A
Company Profits

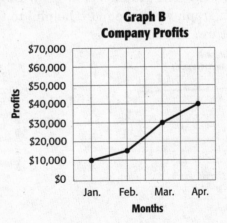

Graph B
Company Profits

1. Which graph would be best to use to convince potential investors to invest in this company?

2. Why might the graph be misleading?

8-9 Practice

Misleading Statistics

ANTIQUES For Exercises 1–3, use the table.

1. Find the mean, median, and mode of the data.

2. Which measure might be misleading in describing the value of each item? Explain.

Antiques	
Item	**Value**
Desk	$150
Table	$850
Painting	$850
Dishes	$750
Sewing Machine	$200

3. Which measure would best describe the value of each item? Explain.

MOUNTAINS For Exercises 4 and 5, use the graph that shows the elevation of the two highest mountain peaks in Alaska.

4. Based on the size of the bars compare the elevations of the mountains.

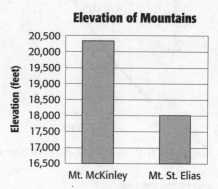

5. Explain how this graph may be misleading.

6. **BODY TEMPERATURE** The graphs below show the hourly body temperature for a hospital patient. Which graph would be more helpful to the doctor in showing the change in body temperature? Explain.

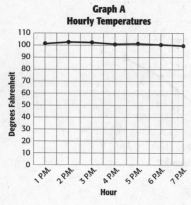

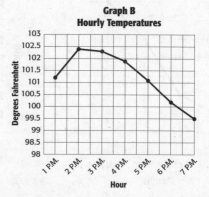

9-1 Study Guide and Intervention

Section Title

The **probability** of a simple event is a ratio that compares the number of favorable outcomes to the number of possible outcomes. Outcomes occur at **random** if each outcome occurs by chance.

Two events that are the only ones that can possibly happen are **complementary events**. The sum of the probabilities of complementary events is 1.

Example 1 What is the probability of rolling a multiple of 3 on a number cube marked with 1, 2, 3, 4, 5, and 6 on its faces.

$$P(\text{multiple of 3}) = \frac{\text{multiples of 3 possible}}{\text{total numbers possible}}$$

$$= \frac{2}{6} \qquad \text{Two numbers are multiples of 3: 3 and 6.}$$

$$= \frac{1}{3} \qquad \text{Simplify.}$$

The probability of rolling a multiple of 3 is $\frac{1}{3}$ or about 33.3%.

Example 2 What is the probability of *not* rolling a multiple of 3 on a number cube marked with 1, 2, 3, 4, 5, and 6 on its faces?

$$P(A) + P(\text{not } A) = 1$$

$$\frac{1}{3} + P(\text{not } A) = 1 \qquad \text{Substitute } \tfrac{1}{3} \text{ for } P(A).$$

$$\underline{-\frac{1}{3} \qquad\qquad -\frac{1}{3}} \qquad \text{Subtract } \tfrac{1}{3} \text{ from each side}$$

$$P(\text{not } A) = \frac{2}{3} \qquad \text{Simplify.}$$

The probability of *not* rolling a multiple of 3 is $\frac{2}{3}$ or about 66.7%.

Exercises

A set of 30 cards is numbered 1, 2, 3, ..., 30. Suppose you pick a card at random without looking. Find the probability of each event. Write as a fraction in simplest form.

1. $P(12)$

2. $P(2 \text{ or } 3)$

3. $P(\text{odd number})$

4. $P(\text{a multiple of 5})$

5. $P(\text{not a multiple of 5})$

6. $P(\text{less than or equal to 10})$

9-1 Practice

Simple Events

A set of cards is numbered 1, 2, 3, ... 24. Suppose you pick a card at random without looking. Find the probability of each event. Write as a fraction in simplest form.

1. $P(5)$ **2.** $P(\text{multiple of 4})$ **3.** $P(6 \text{ or } 17)$

4. $P(\text{not equal to 15})$ **5.** $P(\text{not a factor of 6})$ **6.** $P(\text{odd number})$

COMMUNITY SERVICE The table shows the students involved in community service. Suppose one student is randomly selected to represent the school at a state-wide awards ceremony. Find the probability of each event. Write as a fraction in simplest form.

7. $P(\text{boy})$ **8.** $P(\text{not 6th grader})$

9. $P(\text{girl})$ **10.** $P(\text{8th grader})$

11. $P(\text{boy or girl})$ **12.** $P(\text{6th or 7th grader})$

13. $P(\text{7th grader})$ **14.** $P(\text{not a 9th grader})$

Community Service	
girls	15
boys	25
6th graders	20
7th graders	8
8th graders	12

MENU A delicatessen serves different menu items, of which 2 are soups, 6 are sandwiches, and 4 are salads. How likely is it for each event to happen if you choose one item at random from the menu? Explain your reasoning.

15. $P(\text{sandwich})$ **16.** $P(\text{not a soup})$ **17.** $P(\text{salad})$

18. NUMBER CUBE What is the probability of rolling an even number or a prime number on a number cube? Write as a fraction in simplest form.

19. CLOSING TIME At a convenience store there is a 25% chance a customer enters the store within one minute of closing time. Describe the complementary event and find its probability.

9-2 Study Guide and Intervention

Sample Spaces

A game in which players of equal skill have an equal chance of winning is a **fair game**. A **tree diagram** or table is used to show all of the possible outcomes, or **sample space**, in a probability experiment.

Example 1 WATCHES **A certain type of watch comes in brown or black and in a small or large size. Find the number of color-size combinations that are possible.**

Make a table to show the sample space. Then give the total number of outcomes.

Color	Size
Brown	Small
Brown	Large
Black	Small
Black	Large

There are four different color and size combinations.

Example 2 CHILDREN **The chance of having either a boy or a girl is 50%. What is the probability of the Smiths having two girls?**

Make a tree diagram to show the sample space. Then find the probability of having two girls.

| Child 1 | Child 2 | Sample Space |

The sample space contains 4 possible outcomes. Only 1 outcome has both children being girls. So, the probability of having two girls is $\frac{1}{4}$.

Exercises

For each situation, make a tree diagram or table to show the sample space. Then give the total number of outcomes.

1. choosing an outfit from a green shirt, blue shirt, or a red shirt, and black pants or blue pants

2. choosing a vowel from the word COUNTING and a consonant from the word PRIME

Lesson 9-2

9-2 Practice

Sample Spaces

For each situation, find the sample space using a table or tree diagram.

1. choosing blue, green, or yellow wall paint with white, beige, or gray curtains

Paint	**Curtains**	**Sample Space**

2. choosing a lunch consisting of a soup, salad, and sandwich from the menu shown in the table.

Soup	**Salad**	**Sandwich**
Tortellini	Caesar	Roast Beef
Lentil	Macaroni	Ham
		Turkey

3. **GAME** Kimiko and Miko are playing a game in which each girl rolls a number cube. If the sum of the numbers is a prime number, then Miko wins. Otherwise Kimiko wins. Find the sample space. Then determine whether the game is fair.

9-3 Study Guide and Intervention

The Fundamental Counting Principle

> If event *M* can occur in *m* ways and is followed by event *N* that can occur in *n* ways, then the event *M* followed by *N* can occur in *m* × *n* ways. This is called the **Fundamental Counting Principle**.

Example 1 CLOTHING Andy has 5 shirts, 3 pairs of pants, and 6 pairs of socks. How many different outfits can Andy choose with a shirt, pair of pants, and pair of socks?

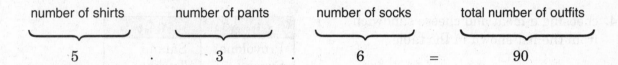

number of shirts		number of pants		number of socks		total number of outfits
5	·	3	·	6	=	90

Andy can choose 90 different outfits.

Exercises

Use the Fundamental Counting Principle to find the total number of outcomes in each situation.

1. rolling two number cubes

2. tossing 3 coins

3. picking one consonant and one vowel

4. choosing one of 3 processor speeds, 2 sizes of memory, and 4 sizes of hard drive

5. choosing a 4-, 6-, or 8-cylinder engine and 2- or 4-wheel drive

6. rolling 2 number cubes and tossing 2 coins

7. choosing a color from 4 colors and a number from 4 to 10

Lesson 9-3

9-3 Practice

The Fundamental Counting Principle

Use the Fundamental Counting Principle to find the total number of outcomes in each situation.

1. choosing from 8 car models, 5 exterior paint colors, and 2 interior colors

2. selecting a year in the last decade and a month of the year

3. picking from 3 theme parks and 1-day, 2-day, 3-day, and 5-day passes

4. choosing a meat and cheese sandwich from the list shown in the table

Cheese	Meat
Provolone	Salami
Swiss	Turkey
American	Tuna
Cheddar	Ham

5. tossing a coin and rolling 2 number cubes

6. selecting coffee in regular or decaf, with or without cream, and with or without sweeteners

7. **COINS** Find the number of possible outcomes if 2 quarters, 4 dimes, and 1 nickel are tossed.

8. **SOCIAL SECURITY** Find the number of possible 9-digit social security numbers if the digits may be repeated.

9. **AIRPORTS** Jolon will be staying with his grandparents for a week. There are four flights that leave the airport near Jolon's home that connect to an airport that has two different flights to his grandparents' hometown. Find the number of possible flights. Then find the probability of taking the earliest flight from each airport if the flight is selected at random.

10. **ANALYZE TABLES** The table shows the kinds of homes offered by a residential builder. If the builder offers a discount on one home at random, find the probability it will be a 4-bedroom home with an open porch. Explain your reasoning.

Number of Bedrooms	Style of Kitchen	Type of Porch
5-bedroom	Mediterranean	Open
4-bedroom	Contemporary	Screen
3-bedroom	Southwestern	
	Colonial	

9-4 Study Guide and Intervention

Permutations

A **permutation** is an arrangement, or listing, of objects in which order is important. You can use the Fundamental Counting Principle to find the number of possible arrangements.

Example 1 Find the value of $5 \cdot 4 \cdot 3 \cdot 2 \cdot 1$.

$5 \cdot 4 \cdot 3 \cdot 2 \cdot 1$
$= 120$ Simplify.

Example 2 Find the value of $4 \cdot 3 \cdot 2 \cdot 1 \cdot 2 \cdot 1$.

$4 \cdot 3 \cdot 2 \cdot 1 \cdot 2 \cdot 1$
$= 48$ Simplify.

Example 3 BOOKS How many ways can 4 different books be arranged on a bookshelf?

This is a permutation. Suppose the books are placed on the shelf from left to right.

There are **4** choices for the first book.
  There are **3** choices that remain for the second book.
    There are **2** choices that remain for the third book.
      There is **1** choice that remains for the fourth book.

$4 \cdot 3 \cdot 2 \cdot 1$
$= 24$ Simplify.

So, there are 24 ways to arrange 4 different books on a bookshelf.

Exercises

Find the value of each expression.

1. $3 \cdot 2 \cdot 1$

2. $7 \cdot 6 \cdot 5 \cdot 4 \cdot 3 \cdot 2 \cdot 1$

3. $6 \cdot 5 \cdot 4 \cdot 3 \cdot 2 \cdot 1 \cdot 3 \cdot 2 \cdot 1$

4. $9 \cdot 8 \cdot 7$

5. How many ways can you arrange the letters in the word GROUP?

6. How many different 4-digit numbers can be created if no digit can be repeated? Remember, a number cannot begin with 0.

Lesson 9-4

9-4 Practice

Permutations

Solve each problem.

1. **NUMBERS** How many different 2-digit numbers can be formed from the digits 4, 6, and 8? Assume no number can be used more than once.

2. **LETTERS** How many permutations are possible of the letters in the word NUMBERS?

3. **PASSENGERS** There are 5 passengers in a car. In how many ways can the passengers sit in the 5 passenger seats of the car?

4. **PAINTINGS** Mr. Bernstein owns 14 paintings, but has only enough wall space in his home to display three of them at any one time: one in the hallway, one in the den, and one in the parlor. How many ways can Mr. Bernstein display three paintings in his home?

5. **DOG SHOW** Mateo is one of the six dog owners in the terrier category. If the owners are selected in a random order to show their dogs, how many ways can the owners show their dogs?

6. **TIME** Michel, Jonathan, and two of their friends each ride their bikes to school. If they have an equally-likely chance of arriving first, what is the probability that Jonathan will arrive first and Michel will arrive second?

7. **BIRTHDAY** Glen received 6 birthday cards. If he is equally likely to read the cards in any order, what is the probability he reads the card from his parents and the card from his sister before the other cards?

CODES For Exercises 8–10, use the following information. A bank gives each new customer a 4-digit code number which allows the new customer to create their own password. The code number is assigned randomly from the digits 1, 3, 5, and 7, and no digit is repeated.

8. What is the probability that the code number for a new customer will begin or end with a 7?

9. What is the probability that the code number will *not* contain a 5?

10. What is the probability that the code number will start with 371?

9-5 Study Guide and Intervention

Combinations

An arrangement, or listing, of objects in which order is *not* important is called a **combination**. You can find the number of combinations of objects by dividing the number of permutations of the entire set by the number of ways each smaller set can be arranged.

Example 1 Jill was asked by her teacher to choose 3 topics from the 8 topics given to her. How many different three-topic groups could she choose?

There are $8 \cdot 7 \cdot 6$ permutations of three-topic groups chosen from eight. There are $3 \cdot 2 \cdot 1$ ways to arrange the groups.

$$\frac{8 \cdot 7 \cdot 6}{3 \cdot 2 \cdot 1} = \frac{336}{6} = 56$$

So, there are 56 different three-topic groups.

Tell whether each situation represents a *permutation* or *combination*. Then solve the problem.

Example 2 On a quiz, you are allowed to answer any 4 out of the 6 questions. How many ways can you choose the questions?

This is a combination because the order of the 4 questions is not important. So, there are $6 \cdot 5 \cdot 4 \cdot 3$ permutations of four questions chosen from six. There are $4 \cdot 3 \cdot 2 \cdot 1$ orders in which these questions can be chosen.

$$\frac{6 \cdot 5 \cdot 4 \cdot 3}{4 \cdot 3 \cdot 2 \cdot 1} = \frac{360}{24} = 15$$

So, there are 15 ways to choose the questions.

Example 3 Five different cars enter a parking lot with only 3 empty spaces. How many ways can these spaces be filled?

This is a permutation because each arrangement of the same 3 cars counts as a distinct arrangement. So, there are $5 \cdot 4 \cdot 3$ or 60 ways the spaces can be filled.

Exercises

Tell whether each situation represents a *permutation* or *combination*. Then solve the problem.

1. How many ways can 4 people be chosen from a group of 11?

2. How many ways can 3 people sit in 4 chairs?

3. How many ways can 2 goldfish be chosen from a tank containing 15 goldfish?

Lesson 9-5

9-5 Practice

Combinations

Solve each problem.

1. **BASKETBALL** In how many ways can a coach select 5 players from a team of 10 players?

2. **BOOKS** In how many ways can 3 books be selected from a shelf of 25 books?

3. **CAFETERIA** In how many ways can you choose 2 side dishes from 15 items?

4. **CHORES** Of 8 household chores, in how many ways can you do three-fourths of them?

5. **ELDERLY** Latanya volunteers to bake and deliver pastries to elderly people in her neighborhood. In how many different ways can Latanya deliver to 2 of the 6 elderly people in her neighborhood?

6. **DELI** A deli makes potato, macaroni, three bean, Caesar, 7-layer, and Greek salads. The deli randomly makes only four salads each day. What is the probability that the four salads made one day are 7-layer, macaroni, Greek, and potato?

7. **AUTOGRAPHS** A sports memorabilia enthusiast collected autographed baseballs from the players in the table. The enthusiast is giving one autographed baseball to each of his three grandchildren. If the baseballs are selected at random, what is the probability that the Hank Aaron, Alex Rodriquez, and Mickey Mantle autographed baseballs are given to his grandchildren?

Player
Cal Ripkin
Hank Aaron
Barry Bonds
Alex Rodriquez
Mickey Mantle

For Exercises 8–10, tell whether each problem represents a *permutation* or a *combination*. Then solve the problem.

8. **LOCKS** In how many ways can three different numbers be selected from 10 numbers to open a keypad lock?

9. **MOVIES** How many ways can 10 DVDs be placed on a shelf?

10. **TRANSPORTATION** Eight people need transportation to the concert. How many different groups of 6 people can ride with Mrs. Johnson?

9-6 Study Guide and Intervention

Problem-Solving Investigation: Act It Out

By acting out a problem, you are able to see all possible solutions to the problem being posed.

Example   CLOTHING Ricardo has two shirts and three pairs of pants to choose from for his outfit to wear on the first day of school. How many different outfits can he make by wearing one shirt and one pair of pants?

Understand   We know that he has two shirts and three pairs of pants to choose from. We can use a coin for the shirts and an equally divided spinner labeled for the pants.

Plan   Let's make a list showing all possible outcomes of tossing a coin and then spinning a spinner.

Solve   H = Heads
T = Tails
Spinner = 1, 2, 3

Flip a Coin	Spin a Spinner
H	1
H	2
H	3
T	1
T	2
T	3

There are six possible outcomes of tossing a coin and spinning a spinner. So, there are 6 different possible outfits that Ricardo can wear for the first day of school.

Check   Tossing a coin has two outcomes and there are two shirts. Spinning a three-section spinner has three outcomes and there are three pairs of pants. Therefore, the solution of 6 different outcomes with a coin and spinner represent the 6 possible outfit outcomes for Ricardo.

Exercises

1. **SCIENCE FAIR** There are 4 students with projects to present at the school science fair. How many different ways can these 4 projects be displayed on four tables in a row?

2. **GENDER** Determine whether tossing a coin is a good way to predict the gender of the next 5 babies born at General Hospital. Justify your answer.

3. **OLYMPICS** Four runners are entered in the first hurdles heat of twelve heats at the Olympics. The first two move on to the next round. Assuming no ties, how many different ways can the four runners come in first and second place?

9-6 Practice

Problem-Solving Investigation: Act It Out

Mixed Problem Solving

For Exercises 1 and 2, use the act it out strategy.

1. **POP QUIZ** Use the information in the table to determine whether tossing a nickel and a dime is a good way to answer a 5-question multiple-choice quiz if each question has answer choices A, B, C, and D. Justify your answer.

Nickel	Dime	Answer Choice
H	H	A
H	T	B
T	H	C
T	T	D

2. **BOWLING** Bill, Lucas, Carmen, and Dena go bowling every week. When ordered from highest to lowest, how many ways can their scores be arranged if Lucas is never first and Carmen always beats Bill?

Use any strategy to solve Exercises 3–6. Some strategies are shown below.

PROBLEM-SOLVING STRATEGIES
• Draw a diagram.
• Use reasonable answers.
• Act it out.

3. **BOOKS** What is the probability of five books being placed in alphabetical order of their titles if randomly put on a book shelf?

4. **NUMBER THEORY** The sum of a 2-digit number and the 2-digit number when the digits are reversed is 77. If the difference of the same two numbers is 45, what are the two 2-digit numbers?

5. **BASEBALL** In one game, Rafael was up to bat 3 times and made 2 hits. In another game, he was up to bat 5 times with no hits. What percent of the times at bat did Rafael make a hit?

6. **RESTAURANT** A restaurant offers the possibility of 168 three-course dinners. Each dinner has an appetizer, an entrée, and a dessert. If the number of appetizers decreases from 7 to 5, find how many fewer possible three-course dinners the restaurant offers.

9-7 Study Guide and Intervention

Theoretical and Experimental Probability

Lesson 9-7

Experimental probability is found using frequencies obtained in an experiment or game. **Theoretical probability** is the expected probability of an event occurring.

Example 1 The graph shows the results of an experiment in which a number cube was rolled 100 times. Find the experimental probability of rolling a 3 for this experiment.

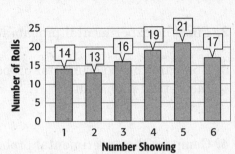

$$P(3) = \frac{\text{number of times 3 occurs}}{\text{number of possible outcomes}}$$

$$= \frac{16}{100} \text{ or } \frac{4}{25}$$

The experimental probability of rolling a 3 is $\frac{4}{25}$, which is close to its theoretical probability of $\frac{1}{6}$.

Example 2 In a telephone poll, 225 people were asked for whom they planned to vote in the race for mayor. What is the experimental probability of Juarez being elected?

Candidate	Number of People
Juarez	75
Davis	67
Abramson	83

Of the 225 people polled, 75 planned to vote for Juarez.

So, the experimental probability is $\frac{75}{225}$ or $\frac{1}{3}$.

Example 3 Suppose 5,700 people vote in the election. How many can be expected to vote for Juarez?

$\frac{1}{3} \cdot 5,700 = 1,900$

About 1,900 will vote for Juarez.

Exercises

For Exercises 1–3, use the graph of a survey of 150 students asked whether they prefer cats or dogs.

1. What is the probability of a student preferring dogs?

2. Suppose 100 students were surveyed. How many can be expected to prefer dogs?

3. Suppose 300 students were surveyed. How many can be expected to prefer cats?

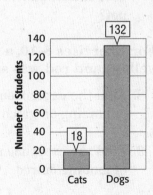

9-7 Practice

Theoretical and Experimental Probability

For Exercises 1–4, a number cube is rolled 24 times and lands on 2 four times and on 6 three times.

1. Find the experimental probability of landing on a 2.

2. Find the experimental probability of *not* landing on a 6.

3. Compare the experimental probability you found in Exercise 1 to its theoretical probability.

4. Compare the experimental probability you found in Exercise 2 to its theoretical probability.

ENTERTAINMENT For Exercises 5–7, use the results of the survey in the table shown.

5. What is the probability that someone in the survey considered reading books or surfing the Internet as the best entertainment value? Write the probability as a fraction.

6. Out of 500 people surveyed, how many would you expect considered reading books or surfing the Internet as the best entertainment value?

Best Entertainment Value	
Type of Entertainment	**Percent**
Playing Interactive Games	48%
Reading Books	22%
Renting Movies	10%
Going to Movie Theaters	10%
Surfing the Internet	9%
Watching Television	1%

7. Out of 300 people surveyed, is it reasonable to expect that 30 considered watching television as the best entertainment value? Why or why not?

For Exercises 8–10, a spinner marked with four sections blue, green, yellow, and red was spun 100 times. The results are shown in the table.

8. Find the experimental probability of landing on green.

9. Find the experimental probability of landing on red.

Section	Frequency
Blue	14
Green	10
Yellow	8
Red	68

10. If the spinner is spun 50 more times, how many of these times would you expect the pointer to land on blue?

9-8 Study Guide and Intervention

Compound Events

A **compound event** consists of two or more simple events. If the outcome of one event does not affect the outcome of a second event, the events are called **independent events**. The probability of two independent events can be found by multiplying the probability of the first event by the probability of the second event.

Example 1 A coin is tossed and a number cube is rolled. Find the probability of tossing tails and rolling a 5.

$P(\text{tails}) = \frac{1}{2}$ $\qquad$ $P(5) = \frac{1}{6}$

$P(\text{tails and 5}) = \frac{1}{2} \cdot \frac{1}{6}$ or $\frac{1}{12}$

So, the probability of tossing tails and rolling a 5 is $\frac{1}{12}$.

Example 2 MARBLES A bag contains 7 blue, 3 green, and 3 red marbles. If Agnes randomly draws two marbles from the bag, replacing the first before drawing the second, what is the probability of drawing a green and then a blue marble?

$P(\text{green}) = \frac{3}{13}$ $\qquad$ 13 marbles, 3 are green

$P(\text{blue}) = \frac{7}{13}$ $\qquad$ 13 marbles, 7 are blue

$P(\text{green, then blue}) = \frac{3}{13} \cdot \frac{7}{13} = \frac{21}{169}$

So, the probability that Agnes will draw a green, then a blue marble is $\frac{21}{169}$.

Exercises

1. Find the probability of rolling a 2 and then an even number on two consecutive rolls of a number cube.

2. A penny and a dime are tossed. What is the probability that the penny lands on heads and the dime lands on tails?

3. Lazlo's sock drawer contains 8 blue and 5 black socks. If he randomly pulls out one sock, what is the probability that he picks a blue sock?

Lesson 9-8

9-8 **Practice**

Compound Events

A number cube is rolled and a spinner like
the one shown is spun. Find each probability.

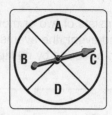

1. $P(6$ and $D)$

2. $P($multiple of 2 and $B)$

3. $P($not 6 and not $A)$

A set of 7 cards is labeled 1–7. A second set of 12 cards contains the following
colors: 3 green, 6 red, 2 blue, and 1 white. One card from each set is selected. Find
each probability.

4. $P(6$ and green$)$

5. $P($prime and blue$)$

6. $P($odd and red$)$

7. $P(7$ and white$)$

8. $P($multiple of 3 and red$)$

9. $P($even and white$)$

A coin is tossed, a number cube is rolled, and a letter is picked from
the word *framer*.

10. $P($tails, 5, $m)$

11. $P($heads, odd, $r)$

12. $P($heads, 6, vowel$)$

13. $P($tails, prime, consonant$)$ **14.** $P($not tails, multiple of 3, $a)$ **15.** $P($not heads, 2, $f)$

16. TOLL ROAD Mr. Espinoza randomly chooses one of five toll booths when
    entering a toll road when driving to work. What is the probability he will
    select the middle toll booth on Monday and Tuesday?

MARBLES For Exercises 17–20, use the
information in the table shown to find each
probability. After a marble is randomly picked
from a bag containing marbles of four different
colors, the color of the marble is observed and
then it is returned to the bag.

Marbles	
Color	Number
White	6
Green	2
Red	1
Blue	3

17. $P($red$)$

18. $P($green, blue$)$

19. $P($red, white, blue$)$

20. $P($blue, blue, blue$)$

10-1 Study Guide and Intervention

Angle Relationships

- An **angle** has two sides that share a common endpoint. The point where the sides meet is called the **vertex**. Angles are measured in **degrees**, where 1 degree is one of 360 equal parts of a circle.
- Angles are classified according to their measure.

| Right Angle | Acute Angle | Obtuse Angle | Straight Angle |

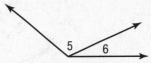

- Two angles are **vertical** if they are opposite angles formed by the intersection of two lines.
- Two angles are **adjacent** if they share a common vertex, a common side, and do not overlap.

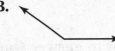

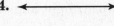

∠1 and ∠3 are vertical angles.   ∠5 and ∠6 are adjacent angles
∠4 and ∠2 are vertical angles.

Example 1 **Classify each angle as** *acute, obtuse, right,* **or** *straight.*

A.   The angle is less than 90°, so it is an acute angle.

B.   The angle is greater than 90°, so it is an obtuse angle.

Example 2 **Label the two angles** *vertical* **or** *adjacent.*

C.   These angles are vertical because they are opposite each other and formed by two intersecting lines.

D. 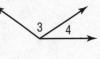 These angles are adjacent because they share a common vertex, a common side, and do not overlap.

Exercises

Classify each angle as *acute, obtuse, right,* **or** *straight.*

1.

2.

3.

4.

Label the angles *vertical* **or** *adjacent.*

5.

6.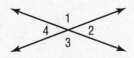

7.

8.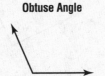

Lesson 10-1

10-1 Practice

Angle Relationships

Name each angle in four ways. Then classify the angle as *acute, right, obtuse*, or *straight*.

1.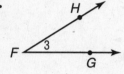

2.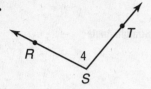

3.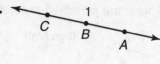

4.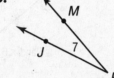

5.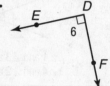

6.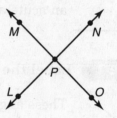

Use the figure at the right to answer Questions 7 and 8.

7. Name two angles that are vertical.

8. Name two angles that are adjacent.

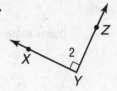

Use the figure at the right to name the following.

9. two acute angles

10. two straight angles

11. two right angles

12. two obtuse angles

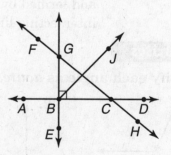

10-2 Study Guide and Intervention
Complementary and Supplementary Angles

- Two angles are **complementary** if the sum of their measure is 90°.

$$m\angle 1 + m\angle 2 = 90°$$

- Two angles are **supplementary** if the sum of their measure is 180°.

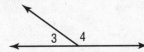

$$m\angle 3 + m\angle 4 = 180°$$

- To **find a missing angle measure,** first determine if the angles are complementary or supplementary. Then write an equation and subtract to find the missing measure.

Example 1 Find the value of *x*.

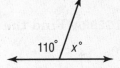

The two angles form a right angle or 90°, so they are complementary,

$$43 + x = 90$$
$$\underline{-43 \qquad\quad -43}$$
$$x = 47$$

Write the equation.
Subtract 43 from each side.

so the value of *x* is 47°.

Example 2 Find the value of *x*.

The two angles form a straight line or 180°, so they are supplementary,

$$110 + x = 180$$
$$\underline{-110 \qquad\quad -110}$$
$$x = \;\;\;70$$

Write the equation.
Subtract 110 from each side.

so the value of *x* is 70°.

Exercises

Find the value of *x* in each figure.

1.

2.

3.

4.

5.

6.

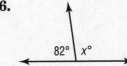

Lesson 10-2

10-2 **Practice**

Complementary and Supplementary Angles

Find the value of x in each figure.

1.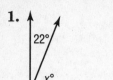

2.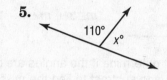

3.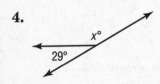

4.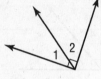

5.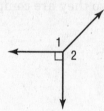

6.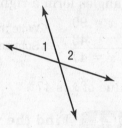

Classify each pair of angles as *complementary*, *supplementary*, or *neither*.

7.

8.

9.

ALGEBRA Find the value of x in each figure.

10.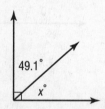

11.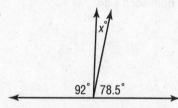

12.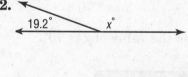

13. **ALGEBRA** If $\angle C$ and $\angle D$ are supplementary, and the measure of $\angle D$ is 45°, what is the measure of $\angle C$?

10-3 Study Guide and Intervention

Statistics: Display Data in a Circle Graph

A graph that shows data as parts of a whole circle is called a **circle graph**. In a circle graph, the percents add up to 100. When percents are not given, you must first determine what part of the whole each item represents.

Example 1 ENERGY Make a circle graph of the data in the table.

Nuclear Reactors in Operation	
Country	Number of Reactors
United States	104
France	59
Japan	54
Other Countries	222

Step 1 Find the total number of reactors: $104 + 59 + 54 + 222 = 439$.

Step 2 Find the ratio that compares each number with the total. Write the ratio as a decimal rounded to the nearest hundredth.

United States: $\frac{104}{439} \approx 0.24$ Japan: $\frac{54}{439} \approx 0.12$

France: $\frac{59}{439} \approx 0.13$ Other: $\frac{222}{439} \approx 0.51$

Step 3 Find the number of degrees for each section of the graph.

United States: $0.24 \cdot 360° \approx 86°$ Japan: $0.12 \cdot 360° \approx 43°$

France: $0.13 \cdot 360° \approx 47°$ Other: $0.51 \cdot 360° \approx 184°$

Step 4 Use a compass to construct a circle and draw a radius. Then use a protractor to draw an 86° angle. This represents the percent of nuclear reactors in the United States.

Step 5 From the new radius, draw a 47° angle for France. Repeat this step for the other two sections. Label each section and give the graph a title.

Nuclear Reactors in Operation, 2001

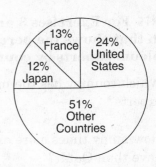

Exercises

1. **SWIMMING** The table shows the number of members of the swim team who competed at the swim meet. Each competed in only one event. Make a circle graph of the data.

Swim Team Member Participation	
Event	Number
Freestyle	18
Breaststroke	7
Backstroke	5
Butterfly	2

Swim Team Member Participation

Lesson 10-3

10-3 Practice

Statistics: Display Data in a Circle Graph

Display each set of data in a circle graph.

1.

Volume of World's Oceans	
Ocean	Percent
Pacific	49%
Atlantic	26%
Indian	21%
Arctic	4%

2.

America's Energy Sources	
Type	Percent
Petroleum	40%
Natural Gas	23%
Coal	22%
Nuclear	8%
Other	7%

Volume of World's Oceans

America's Energy Sources

EXPORTS For Exercises 3 and 4, use the circle graph that shows the percent of Persian Gulf petroleum exports by country.

Persian Gulf Exports

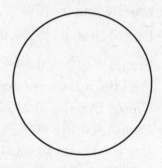

3. Which country has the most petroleum exports?

4. How many times more exports does Iran have than Qatar?

DATA SENSE For each graph, find the missing values.

5. Recycled Products

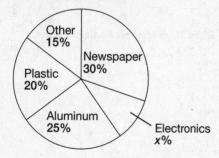

6. Time Management

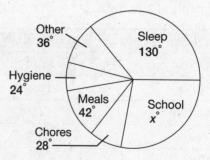

10-4 Study Guide and Intervention

Triangles

A **triangle** is a figure with three sides and three angles. The symbol for triangle is △. The sum of the measures of the angles of a triangle is 180°. You can use this to find a missing angle measure in a triangle.

Example 1 Find the value of *x* in △*ABC*.

$x + 66 + 52 = 180$ The sum of the measures is 180.
$x + 118 = 180$ Simplify.
$\underline{\quad - 118 \quad - 118}$ Subtract 118 from each side.
$x = 62$

The missing angle is 62°.

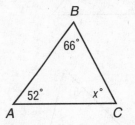

Triangles can be classified by the measures of their angles. An **acute triangle** has three acute angles. An **obtuse triangle** has one obtuse angle. A **right triangle** has one right angle.

Triangles can also be classified by the lengths of their sides. Sides that are the same length are **congruent segments** and are often marked by tick marks. In a **scalene triangle**, all sides have different lengths. An **isosceles** triangle has at least two congruent sides. An **equilateral triangle** has all three sides congruent.

Example 2 Classify the triangle by its angles and by its sides.

The triangle has one obtuse angle and two sides the same length. So, it is an obtuse, isosceles triangle.

Exercises

Find the missing measure in each triangle. Then classify the triangle as *acute*, *right*, or *obtuse*.

1.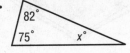

2.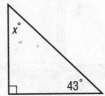

3.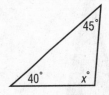

Classify each triangle by its angles and by its sides.

4.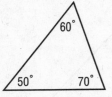

5.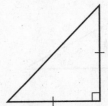

6.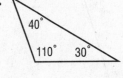

10-4 Practice

Triangles

Find the value of x.

1.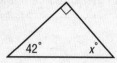

2.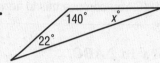

3.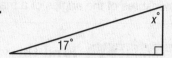

4.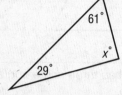

5.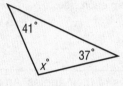

6.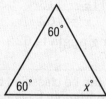

Find the missing measure in each triangle with the given angle measures.

7. $45°, 35.8°, x°$

8. $100°, x°, 40.7°$

9. $x°, 90°, 16.5°$

10. Find the third angle of a right triangle if one of the angles measures $24°$.

11. What is the third angle of a right triangle if one of the angles measures $51.1°$?

12. ALGEBRA Find $m\angle A$ in $\triangle ABC$ if $m\angle B = 38°$ and $m\angle C = 38°$.

13. ALGEBRA In $\triangle XYZ$, $m\angle Z = 113°$ and $m\angle X = 28°$. What is $m\angle Y$?

Classify the marked triangle in each object by its angles and by its sides.

14.

15.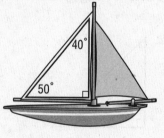

16.

ALGEBRA Find the value of x in each triangle.

20.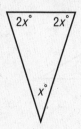

21.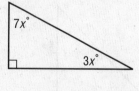

22.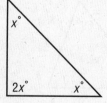

10-5 Study Guide and Intervention

Problem-Solving Investigation: Use Logical Reasoning

Lesson 10-5

Logical reasoning is a method of problem solving that uses **inductive reasoning,** making a rule after seeing several examples, or **deducting reasoning,** using a rule to make a decision.

Example Use the formula $d = rt$ where d is distance, r is rate, and t is time to determine how far a car will travel after 4 hours if it is traveling at a constant rate of 65 miles per hour.

Understand   You know the car will travel for 4 hours at a constant rate of 65 mi/h.

Plan   Try a few examples to find a pattern. Make a table.

Solve

Seconds Passed	Distance Traveled
1	65
1.5	97.5
2	130
2.5	162.5
3	195
t	$65t$

After each hour, the car will travel 65 miles. So, after 4 hours the car will travel 260 mi.

Check   The formula is $d = rt$ so $d = 65 \times 4$ or 260 mi.

Exercises

Solve the following problems using logical reasoning.

1. **TRAVEL** Use the formula $d = rt$ where d is the distance, r is the rate, and t is the time to determine how far the Moralez family has traveled if they are driving at a rate of 72 miles per hour for 9 hours.

2. **CELL PHONES** Determine the cost per phone call if Maria made 30 calls last month and her total bill for the month was $45.00.

3. **MUSIC** Sarah, Juan, and Derrick play the piano, trumpet, and violin, but not necessarily in that order. Sarah and Derrick sit on either side of the trumpet player. Sarah does not play the violin. Who plays the violin?

10-5 Practice

Problem-Solving Investigation: Use Logical Reasoning

Mixed Problem Solving

For Exercises 1 and 2, use logical reasoning to solve the problem.

1. **TOWNS** Tia, Bianca, and Hiroko live in the towns of Parkside, Westlake, and Summerville, but not necessarily in that order. Tia and her friend that lives in Westlake helped Bianca with her chores. Bianca does not live in Parkside. Where does Tia live? Did you use inductive or deductive reasoning?

2. **GEOMETRY** Draw a right triangle. Mark the midpoints of each side of the triangle and draw a smaller triangle by connecting the midpoints. Do this several more times. What can you conclude about the smaller triangle? Did you use inductive or deductive reasoning?

Use any strategy to solve Exercises 3–6. Some strategies are shown below.

PROBLEM-SOLVING STRATEGIES
• Look for a pattern.
• Use a graph.
• Use logical reasoning.

3. **ANGLES** One angle of a triangle is 33° less than the other two angles. Find the measures of the angles of the triangle. Did you use inductive or deductive reasoning?

4. **METEORITES** An astronomer found three meteorites weighing 9.4 pounds, 5.7 pounds, and 24.5 pounds. If 1 kilogram weighs 2.2 pounds, find the average mass of the meteorites in kilograms.

5. **PUBLIC TRANSPORTATION** A bus stopped at a bus stop and 12 people got on and 5 got off. At the next stop, 14 people got on and 3 got off. If the number of passengers has doubled, find the number of passengers on the bus.

6. **DISCOUNTS** The table shows the different discounts two stores offer for the same product. Which store offers the better price after the discount is applied and by how much?

	Price	Discount
Store A	$129.00	$25
Store B	$139.00	25%

10-6 Study Guide and Intervention
Quadrilaterals

Quadrilaterals can be classified using their angles and sides. The best description of a quadrilateral is the one that is the most specific.

			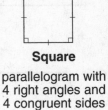	
Trapezoid one pair of parallel sides	**Parallelogram** opposite sides parallel and opposite sides congruent	**Rectangle** parallelogram with 4 right angles	**Rhombus** parallelogram with 4 congruent sides	**Square** parallelogram with 4 right angles and 4 congruent sides

Examples Classify the quadrilateral using the name that *best* describes it.

1    The quadrilateral is a parallelogram with 4 congruent sides. It is a rhombus.

2 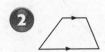   The quadrilateral has one pair of parallel sides. It is a trapezoid.

3 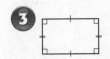   The quadrilateral is a parallelogram with 4 right angles. It is a rectangle.

Example Find the missing measure in the quadrilateral.

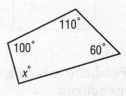

$$100 + 110 + 6 + x = 360 \quad \text{Write the equation.}$$
$$270 + x = 360 \quad \text{Simplify.}$$
$$\underline{-270 \qquad -270} \quad \text{Subtract 270 from}$$
$$x = 90 \quad \text{each side.}$$

So, the missing measure is 90°.

Exercises

Classify the quadrilateral using the name that *best* describes it.

1.

2.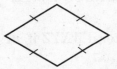

3.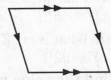

Find the missing angle measure in each quadrilateral.

4.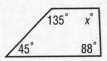

5.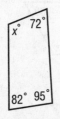

6.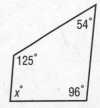

Lesson 10-6

10-6 **Practice**

Quadrilaterals

Classify each quadrilateral using the name that best describes it.

1.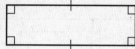

2.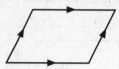

3.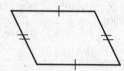

4.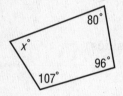

5.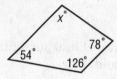

6.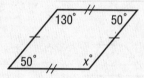

ALGEBRA Find the missing angle measure in each quadrilateral.

7.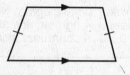
80°
x°
107°
96°

8.
x°
54°
78°
126°

9.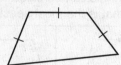
130°   50°
50°   x°

10.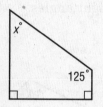
x°
125°

11.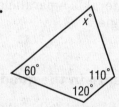
x°
60°
110°
120°

12.
152°
x°

Find the missing measure in each quadrilateral with the given angle measures.

13. 63.2°, 56°, 111.7°, x°

14. 31.7°, x°, 161.3°, 51.4°

15. x°, 122.4°, 53.7°, 90°

16. 83.7°, 137.2°, x°, 28.5°

17. **ALGEBRA** Find $m\angle C$ in quadrilateral $ABCD$ if $m\angle A = 110°$, $m\angle B = 88°$, and $m\angle D = 55°$.

18. **ALGEBRA** What is $m\angle Z$ in quadrilateral $WXYZ$ if $m\angle W = 86°$, $m\angle X = 88°$, and $m\angle Y = 92°$?

ALGEBRA Find the value of x in each quadrilateral.

19.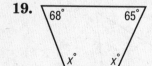
68°   65°
x°   x°

20.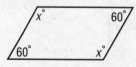
x°   60°
60°   x°

21.
3x°   3x°
3x°   3x°

10-7　Study Guide and Intervention

Similar Figures

Figures that have the same shape but not necessarily the same size are *similar figures*. The symbol ~ means *is similar to*. You can use proportions to find the missing length of a side in a pair of similar figures.

For example △*ABC* ~ △*DEF*.

Corresponding angles

∠ *A* ≅ ∠ *D*

∠ *B* ≅ ∠ *E*

∠ *C* ≅ ∠ *F*

Corresponding sides

$$\frac{5}{10} = \frac{4}{8} = \frac{3}{6}$$

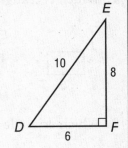

Example 1　If *MNOP* ~ *RSTU*, find the length of $\overline{ST}$.

Since the two figures are similar, the ratios of their corresponding sides are equal. You can write and solve a proportion to find $\overline{ST}$.

$\dfrac{PO}{UT} = \dfrac{NO}{ST}$　　　Write a proportion.

$\dfrac{7}{28} = \dfrac{5}{n}$　　　Let *n* represent the length of *ST*. Then substitute.

$7n = 28(5)$　　　Find the cross products.

$7n = 140$　　　Simplify.

$n = 20$　　　Divide each side by 7.

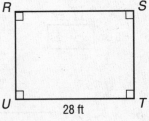

The length of $\overline{ST}$ is 20 feet.

Exercises

Find the value of *x* in each pair of similar figures.

1.

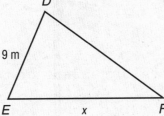

2.

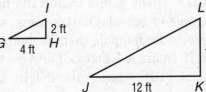

3.

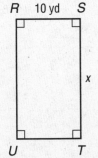

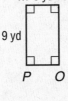

4.

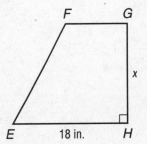

Lesson 10-7

10-7 Practice

Similar Figures

1. Which rectangle is similar to rectangle *RSTU*?

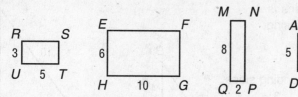

2. Which triangle is similar to triangle *XYZ*?

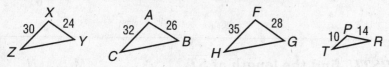

Find the value of *x* in each pair of similar figures.

3.

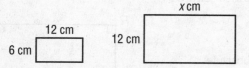

4.

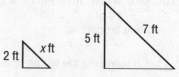

5.

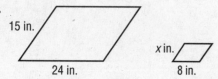

6.

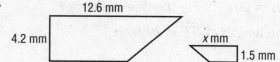

7. FLAGPOLES Tasha wants to find the height of the flagpole at school. One morning, she determines the flagpole casts a shadow of 12 feet. If Tasha is 5 feet tall and casts a shadow of 3 feet, what is the height of the flagpole?

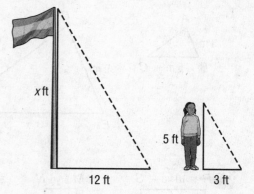

10-8 Study Guide and Intervention

Polygons and Tessellations

A **polygon** is a simple, closed figure formed by three or more straight lines. A simple figure does not have lines that cross each other. You have drawn a closed figure when your pencil ends up where it started. Polygons can be classified by the number of sides they have.

pentagon
5 sides

hexagon
6 sides

heptagon
7 sides

octagon
8 sides

nonagon
9 sides

decagon
10 sides

A polygon that has all sides congruent and all angles congruent is called a **regular polygon**.

Examples Determine whether each figure is a polygon. If it is, classify the polygon and state whether it is regular. If it is *not* a polygon, explain why.

1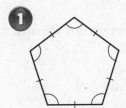

The figure has 5 congruent sides and 5 congruent angles. It is a regular pentagon.

2

The figure is not a polygon because it has sides that overlap.

Exercises

Determine whether each figure is a polygon. If it is, classify the polygon and state whether it is regular. If it is *not* a polygon, explain why.

1.
120°

2.

3.

4.

5.

6.

Lesson 10-8

10-8 Practice

Polygons and Tessellations

Determine whether each figure is a polygon. If it is, classify the polygon and state whether it is regular. If it is not a polygon, explain why.

1.

2.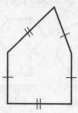

3.

4.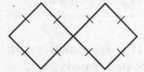

5.

6.

Find the measure of an angle in each polygon if the polygon is regular. Round to the nearest tenth of a degree if necessary.

7. dodecagon (12-sided)

8. 14-gon

9. 18-gon

10. 36-gon

Classify the polygons that are used to create each tessellation.

11.

12.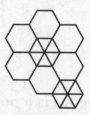

13. What is the perimeter of a regular decagon with sides 6.2 meters long?

14. Find the perimeter of a regular hexagon having sides $5\frac{2}{3}$ inches long.

KITES For Exercises 15–17, use the following information. A kite manufacturer makes kites in the shape of the figure shown.

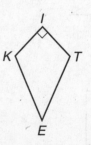

15. Classify the shape of the kite.

16. If $\angle K \cong \angle T$ and $\angle E = 30°$, find $m\angle K$ and $m\angle T$.

17. Can a tessellation be made by using the shape of the kite? Justify your answer.

10-9 Study Guide and Intervention

Translations

A **translation** is the movement of a geometric figure in some direction without turning the figure. When translating a figure, every point of the original figure is moved the same distance and in the same direction. To graph a translation of a figure, move each vertex of the figure in the given direction. Then connect the new vertices.

Example Triangle ABC has vertices $A(-4, -2)$, $B(-2, 0)$, and $C(-1, -3)$. Find the vertices of triangle $A'B'C'$ after a translation of 5 units right and 2 units up.

Add 5 to each x-coordinate.   Add 2 to each y-coordinate.

Vertices of $\triangle ABC$	$(x + 5, y + 2)$	Vertices of $\triangle A'B'C'$
$A(-4, -2)$	$(-4 + 5, -2 + 2)$	$A'(1, 0)$
$B(-2, 0)$	$(-2 + 5, 0 + 2)$	$B'(3, 2)$
$C(-1, -3)$	$(-1 + 5, -3 + 2)$	$C'(4, -1)$

The coordinates of the vertices of $\triangle A'B'C'$ are $A'(1, 0)$, $B'(3, 2)$, and $C'(4, -1)$.

Exercises

1. Translate $\triangle GHI$ 1 unit left and 5 units down.

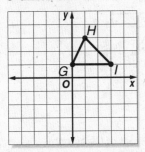

2. Translate rectangle $LMNO$ 4 units right and 3 units up.

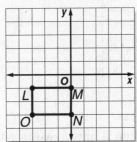

Triangle RST has vertices $R(3, 2)$, $S(4, -2)$, and $T(1, -1)$. Find the vertices of $R'S'T'$ after each translation. Then graph the figure and its translated image.

3. 5 units left, 1 unit up

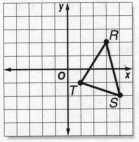

4. 3 units left, 2 units down

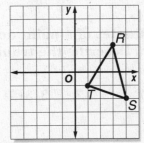

Lesson 10-9

10-9 Practice

Translations

1. Translate rectangle *ABCD* 3 units right and 4 units down. Graph rectangle *A′B′C′D′*.

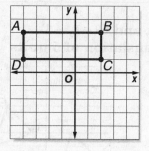

2. Triangle *PQR* is translated 3 units left and 3 units down. Then the translated figure is translated 6 units right. Graph the resulting triangle.

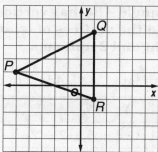

Triangle *EFG* has vertices *E*(1, 1), *F*(4, −3), and *G*(−2, 0). Find the vertices of *E′F′G′* after each translation. Then graph the figure and its translated image.

3. 3 units left, 2 units down

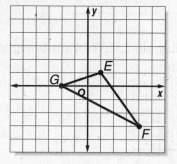

4. 4 units up

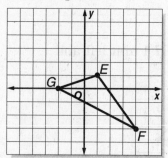

5. SEATS Jatin was given a new seating assignment in science class. The diagram shows his old seat and his new seat. Describe this translation in words and as an ordered pair.

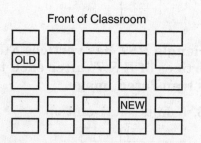

REASONING The coordinates of a point and its image after a translation are given. Describe the translation in words and as an ordered pair.

6. $A(1, -2) \rightarrow A'(3, 4)$

7. $H(3, 3) \rightarrow H'(-4, 0)$

8. $Z(-2, -4) \rightarrow Z'(1, -5)$

10-10 Study Guide and Intervention
Reflections

Figures that match exactly when folded in half have **line symmetry**. Each fold line is called a **line of symmetry**. Some figures have more than one line of symmetry.

Examples Determine whether each figure has line symmetry. If so, draw all lines of symmetry.

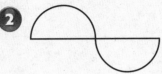 no symmetry

A type of transformation where a figure is flipped over a line of symmetry is a **reflection**. To draw the reflection of a polygon, find the distance from each vertex of the polygon to the line of symmetry. Plot the new vertices the same distance from the line of symmetry but on the other side of the line. Then connect the new vertices to complete the reflected image.

Example 3 Triangle *DEF* has vertices *D*(2, 2), *E*(5, 4), and *F*(1, 5). Find the coordinates of the reflected image. Graph the figure and its reflected image over the *x*-axis.

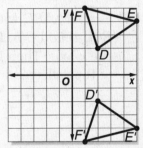

Plot the vertices and connect to form △*DEF*. The *x*-axis is the line of symmetry. The distance from a point on △*DEF* to the line of symmetry is the same as the distance from the line of symmetry to the reflected image. The image coordinates are *D'*(2, −2), *E'*(5, −4), and *F'*(1, −5).

Exercises

For Exercises 1 and 2, determine which figures have line symmetry. Write *yes* or *no*. If *yes*, draw all lines of symmetry.

1.

2.

3. Triangle *ABC* has vertices *A*(0, 4), *B*(2, 1), and *C*(4, 3). Find the coordinates of the vertices of *ABC* after a reflection over the *x*-axis. Then graph the figure and its reflected image.

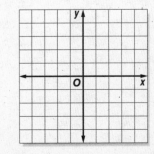

10-10 Practice

Reflections

Determine whether each figure has line symmetry. If so, copy the figure and draw all lines of symmetry.

1.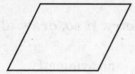

2.

3.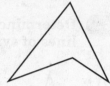

4.

5.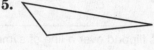

6.

7. Graph △*ABC* with vertices *A*(2, 2), *B*(5, 4),and *C*(5, 1) and its reflection over the *x*-axis. Then find the coordinates of the reflected image.

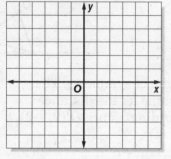

8. Graph square *ABCD* with vertices *A*(−1, 2), *B*(2, −1), *C*(5, 2), and *D*(2, 5) and its reflection over the y-axis. Then find the coordinates of the reflected image.

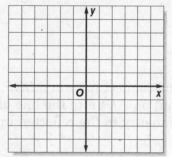

The coordinates of a point and its image after a reflection are given. Describe the reflection as over the *x*-axis or *y*-axis.

9. $B(1, -2) \rightarrow B'(1, 2)$

10. $J(-3, 5) \rightarrow J'(-3, -5)$

11. $W(-7, -4) \rightarrow W'(7, -4)$

11-1 Study Guide and Intervention

Area of Parallelograms

The area A of a parallelogram equals the product of its base b and its height h.

$$A = bh$$

The **base** is any side of a parallelogram.

h

The **height** is the length of the segment perpendicular to the base with endpoints on opposite sides.

b

Example 1 Find the area of a parallelogram if the base is 6 inches and the height is 3.7 inches.

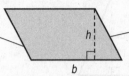

3.7 in.

6 in.

Estimate $A = 6 \cdot 4$ or 24 in^2

$A = bh$ Area of a parallelogram

$A = 6 \cdot 3.7$ Replace b with 6 and h with 3.7.

$A = 22.2$ Multiply.

The area of the parallelogram is 22.2 square inches. This is close to the estimate.

Example 2 Find the area of the parallelogram at the right.

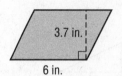

12 cm

8 cm

Estimate $A = 10 \cdot 10$ or 100 cm^2

$A = bh$ Area of a parallelogram

$A = 12 \cdot 8$ Replace b with 12 and h with 8.

$A = 96$ Multiply.

The area of the parallelogram is 96 square centimeters. This is close to the estimate.

Exercises

Find the area of each parallelogram. Round to the nearest tenth if necessary.

1.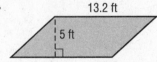
13.2 ft

5 ft

2.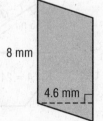
8 mm

4.6 mm

3.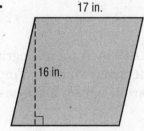
17 in.

16 in.

Lesson 11-1

11-1 Practice

Area of Parallelograms

Find the area of each parallelogram. Round to the nearest tenth if necessary.

1.
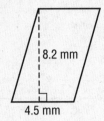
8.2 mm
4.5 mm

2.
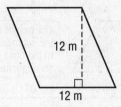
12 m
12 m

3.
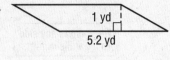
1 yd
5.2 yd

4.
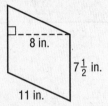
8 in.
$7\frac{1}{2}$ in.
11 in.

5.
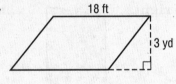
0.7 cm   0.8 cm
0.9 cm

6.
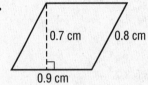
14 ft   13 ft
$15\frac{1}{4}$ ft

7.
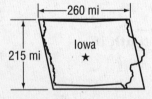
18 ft
3 yd

8.
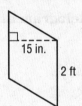
15 in.
2 ft

9.
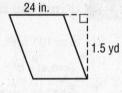
24 in.
1.5 yd

GEOGRAPHY Estimate the area of each state.

10.
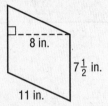
← 260 mi →
Iowa ★
215 mi

11.

135 mi   New Jersey ★
55 mi

12. ALGEBRA A parallelogram has an area of 240 square meters. Find the height of the parallelogram if the base is 20 meters.

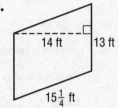
$A = 240 \text{ m}^2$ h m
20 m

13. ALGEBRA What is the base of a parallelogram if the height is 5 feet and the area is 65 square feet?

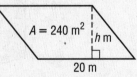
$A = 65 \text{ ft}^2$ 5 ft
b ft

11-2 Study Guide and Intervention

Area of Triangles and Trapezoids

The area A of a triangle equals half the product of its base b and its height h.

$A = \frac{1}{2}bh$

The **base** of a triangle can be any of its sides.

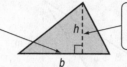

The **height** is the distance from a base to the opposite vertex.

h

b

A trapezoid has two bases, b_1 and b_2. The height of a trapezoid is the distance between the two bases. The area A of a trapezoid equals half the product of the height h and the sum of the bases b_1 and b_2.

$$A = \frac{1}{2}h(b_1 + b_2)$$

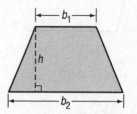

Example 1 **Find the area of the triangle.**

Estimate $\frac{1}{2}(6)(5) = 15$

$A = \frac{1}{2}bh$ Area of a triangle

$A = \frac{1}{2} \cdot 6 \cdot 4.5$ Replace b with 6 and h with 4.5.

$A = 13.5$ Multiply.

The area of the triangle is 13.5 square inches. This is close to the estimate.

4.5 in.

6 in.

Example 2 **Find the area of the trapezoid.**

$A = \frac{1}{2}h(b_1 + b_2)$ Area of a trapezoid

$A = \frac{1}{2}(4)(3 + 6)$ Replace h with 4, b_1 with 3, and b_2 with 6.

$A = 18$ Simplify.

The area of the trapezoid is 18 square centimeters.

3 cm

4 cm

6 cm

Exercises

Find the area of each figure. Round to the nearest tenth if necessary.

1.

7 ft
12 ft

2.
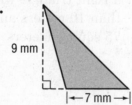
9 mm
←7 mm→|

3.
|←——— 14 in. ———→|
5 in.
7 in.

4.
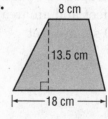
8 cm
13.5 cm
|←——— 18 cm ———→|

Lesson 11-2

11-2 Practice

Area of Triangles and Trapezoids

Find the area of each figure. Round to the nearest tenth if necessary.

1.
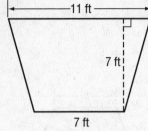

2.
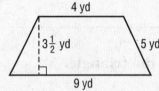

3.
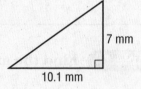

4.
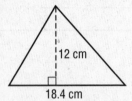

5.
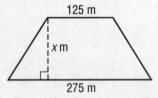

6.
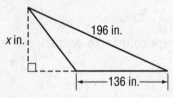

7. GEOGRAPHY The shape of Arkansas is roughly trapezoidal with bases of 150 miles and 250 miles and a height of 260 miles. What is the approximate area of Arkansas?

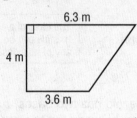

ALGEBRA Find the height of each figure.

8. Area = 23,000 m²

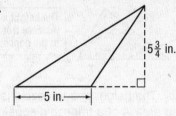

9. Area = 6,460 in²

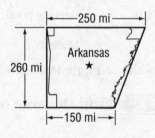

Draw and label each figure. Then find the area.

10. a trapezoid with a height less than 5 feet and an area greater than 50 square feet

11. a right triangle with a base greater than 10 meters and an area greater than 75 square meters

11-3 Study Guide and Intervention

Circles and Circumference

A **circle** is the set of all points in a plane that are the same distance from a given point, called the **center**. The **diameter** d is the distance across the circle through its center. The **radius** r is the distance from the center to any point on the circle. The **circumference** C is the distance around the circle. The circumference C of a circle is equal to its diameter d times π, or 2 times its radius r times π.

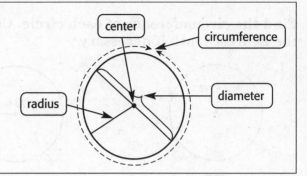

Example 1 Find the circumference of a circle with a diameter of 7.5 centimeters.

$C = \pi d$ Circumference of a circle.

$C \approx 3.14 \times 7.5$ Replace π with 3.14 and d with 7.5.

$C \approx 23.55$ The circumference of the circle is about 23.55 centimeters.

Example 2 If the radius of a circle is 14 inches, what is its circumference?

$C = 2\pi r$

$C \approx 2 \times 3.14 \times 14$ Replace π with 3.14 and r with 14.

$C \approx 87.92$ The circumference of the circle is about 87.92 inches.

Exercises

Find the circumference of each circle. Use 3.14 or $\dfrac{22}{7}$ for π. Round to the nearest tenth if necessary.

1.
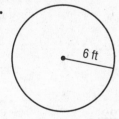
6 ft

2.
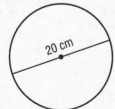
20 cm

3.
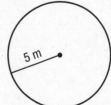
5 m

4.
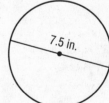
7.5 in.

5. diameter = 15 km

6. radius = 21 mi

7. radius = 50 m

8. diameter = 600 ft

9. radius = 62 mm

10. diameter = 7 km

11. radius = 95 in.

12. diameter = 6.3 m

13. diameter = $5\dfrac{1}{4}$ cm

Lesson 11-3

11-3 Practice

Circles and Circumference

Find the circumference of each circle. Use 3.14 or $\frac{22}{7}$ for π. Round to the nearest tenth if necessary.

1.
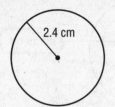
2.4 cm

2.
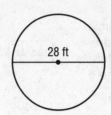
28 ft

3.
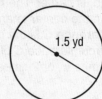
1.5 yd

4.

4.2 mm

5.
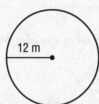
12 m

6.
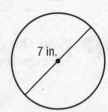
7 in.

7. radius = $2\frac{1}{3}$ ft

8. radius = 11.9 m

9. diameter = $5\frac{5}{6}$ mi

10. radius = $6\frac{1}{8}$ in.

11. diameter = $17\frac{1}{2}$ft

12. radius = 9.2 km

Estimate to find the approximate circumference of each circle. Explain which approximation of π you used.

13.
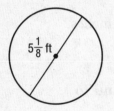
$5\frac{1}{8}$ ft

14.
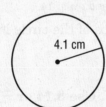
4.1 cm

15.
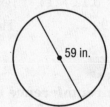
59 in.

ALGEBRA Find the diameter or radius of each circle. Use 3.14 or $\frac{22}{7}$ for π. Round to the nearest tenth if necessary.

16. $C = 32$ m, diameter = ___

17. $C = 55$ mi, radius = ___

18. HELICOPTERS The landing circle for helicopters on the roof of a hospital has a radius of 20 yards. To the nearest yard, find its circumference.

19. SPA A circular spa has a diameter of 12 feet. The spa is decorated with 4-inch porcelain tiles around the rim. How many tiles surround the rim of the spa? Round to the nearest whole tile.

11-4 Study Guide and Intervention

Area of Circles

The area A of a circle equals the product of pi (π) and the square of its radius r.

$$A = \pi r^2$$

Example 1 **Find the area of the circle.**

$A = \pi r^2$ Area of circle

$A \approx 3.14 \cdot 5^2$ Replace π with 3.14 and r with 5.

$A \approx 78.5$

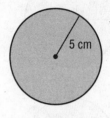

5 cm

The area of the circle is approximately 78.5 square centimeters.

Example 2 **Find the area of a circle that has a diameter of 9.4 millimeters.**

$A = \pi r^2$ Area of a circle

$A \approx 3.14 \cdot 4.7^2$ Replace π with 3.14 and r with 9.4 ÷ 2 or 4.7.

$A \approx 69.4$

The area of the circle is approximately 69.4 square millimeters.

Exercises

Find the area of each circle. Use 3.14 for π. Round to the nearest tenth.

1.
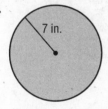
7 in.

2.
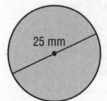
25 mm

3.
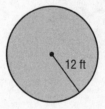
12 ft

4. radius = 2.6 cm

5. radius = 14.3 in.

6. diameter = $5\frac{1}{2}$ yd

7. diameter = $4\frac{3}{4}$ mi

8. diameter = 7.9 mm

9. radius = $2\frac{1}{5}$ ft

Lesson 11-4

11-4 Practice

Area of Circles

Find the area of each circle. Use 3.14 for π.
Round to the nearest tenth if necessary.

1.

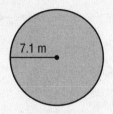

2.

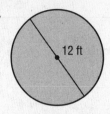

3.

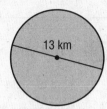

4.

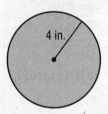

5.

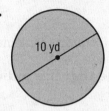

6.

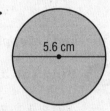

7. diameter = 9.4 mm

8. diameter = $3\frac{1}{2}$ ft

9. radius = $6\frac{1}{4}$ in.

10. radius = $4\frac{3}{4}$ yd

11. diameter = $15\frac{1}{2}$ mi

12. radius = 7.9 km

Estimate to find the approximate area of each circle.

13.

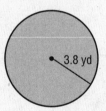

14.

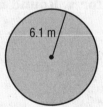

15.

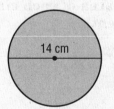

16. **SPOTLIGHT** A spotlight can be adjusted to effectively light a circular area of up to 6 meters in diameter. To the nearest tenth, what is the maximum area that can be effectively lit by the spotlight?

17. **ARCHERY** The bull's eye on an archery target has a radius of 3 inches. The entire target has a radius of 9 inches. To the nearest tenth, find the area of the target outside of the bull's eye.

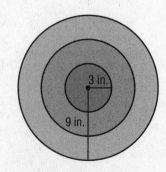

11-5 Study Guide and Intervention

Problem-Solving Investigation: Solve a Simpler Problem

When problem solving, sometimes it is easier to **solve a simpler problem** first to find the correct strategy for solving a more difficult problem.

Example SPORTS West High School wants to paint field blue, but not the center. The diagram below shows the dimensions of the field and center circle. How much area will they need to paint blue?

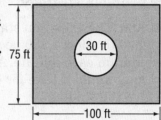

Understand	You know that the field is one large rectangle and the center symbol is a large circle.
Plan	You can find the area of the rectangle and the area of the circle and subtract.

Solve	Area of rectangle: $A = \ell \times w$
	                                    $A = 100 \times 75$ or 7500
	
	 Area if circle: $A = \pi r^2$
	                                    $A = 3.14 \times 15^2$ or 706.5
	
	 Subtract: $7500 - 706.5$ or 6793.5 ft^2
	
	 So, they would need to paint 6,793.5 square feet of field.

Check	Use estimation to check. The area of the entire field is 7,500ft and the circle is approximately 700 feet so the area should be less than 6,800 feet. Since 6,793.5 is less than 6,800ft, the answer is reasonable.

Exercises

1. **FRAMES** Joan wants to paint her favorite picture frame. How much paint would she need to use in order to cover just the frame?

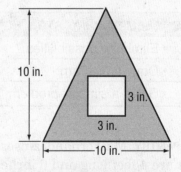

2. **WALLPAPER** Richard wants to wallpaper one wall of his bathroom. He has two semi-circular windows along the wall. How much wallpaper must he purchase?

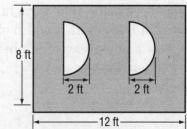

Lesson 11-5

11-5 Practice

Problem-Solving Investigation: Solve a Simpler Problem

Mixed Problem Solving

Solve Exercises 1 and 2. Use the solve a simpler problem strategy.

1. **STADIUM** The exits in a stadium are designed to allow 1,200 people to leave the stadium each minute. At this rate, how long would it take for 10,800 people to leave the stadium?

2. **PHARMACY** A city has three major pharmacy chains which have a total of 895,000 customers. Approximately how many customers do business at each major pharmacy?

Pharmacy	Percent
A	54.8%
B	32.4%
C	12.8%

Use any strategy to solve Exercises 3-6. Some strategies are shown below.

PROBLEM-SOLVING STRATEGIES
• Eliminate possibilities.
• Draw a diagram.
• Solve a simpler problem.

3. **CARPENTRY** Mr. Fernandez uses 7 boards that are 4 feet long and 6 inches wide to make one bookshelf. If he buys lumber in lengths of 8 feet with a width of 12 inches, how many pieces of lumber does he need to purchase to make 5 bookshelves?

4. **AREA** Stacey is making a stained glass window above her front doorway in the shape as shown in the figure. To the nearest tenth, what is the area of the shaded portion of the window?

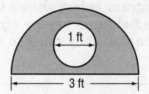

5. **QUALITY CONTROL** For every 250 televisions tested, 3 televisions are found to be defective. How many televisions were tested if 48 televisions were found defective?

6. **APPLIANCE REPAIR** An appliance repair company charged $35 to make a house call. After arriving, the company charged $10 for every 15 minutes of labor. How much was the repair bill if the new parts cost $23 and the appliance took 45 minutes to repair?

11-6 Study Guide and Intervention

Area of Composite Figures

Lesson 11-6

> **Composite figures** are made of triangles, quadrilaterals, semicircles, and other two-dimensional figures. To find the area of a composite figure, separate it into figures whose areas you know how to find, and then add the areas.

Example 1 Find the area of the figure at the right in square feet.

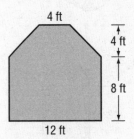

The figure can be separated into a rectangle and a trapezoid. Find the area of each.

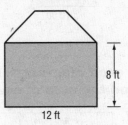

Area of Rectangle

$A = \ell w$ Area of a rectangle

$A = 12 \cdot 8$ Replace ℓ with 12 and w with 8.

$A = 96$ Multiply.

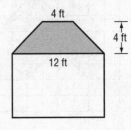

Area of Trapezoid

$A = \frac{1}{2}h(b_1 + b_2)$ Area of a trapezoid

$A = \frac{1}{2}(4)(4 + 12)$ Replace h with 4, b_1 with 4, and b_2 with 12.

$A = 32$ Multiply.

The area of the figure is 96 + 32 or 128 square feet.

Exercises

Find the area of each figure. Use 3.14 for π. Round to the nearest tenth if necessary.

1.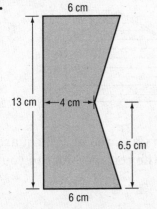

2.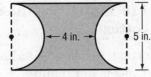

3.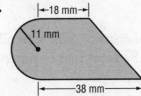

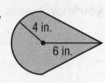

11-6 **Practice**

Area of Composite Figures

Find the area of each figure. Round to the nearest tenth if necessary.

1.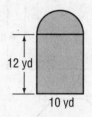
4 in.
6 in.

2.
5.5 ft
6 ft
8 ft

3.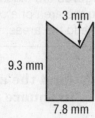
3 mm
9.3 mm
7.8 mm

4.
12 yd
10 yd

5.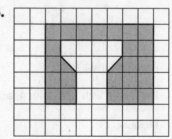
12 cm
8 cm   9 cm
22 cm

6.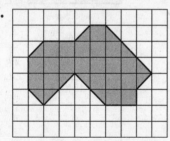
1.8 m
3.2 m   4.5 m
4.5 m
4.5 m
6.5 m

In each diagram below, one square unit represents 5 square meters. Find the area of each figure.

7.

8.

9. AUDITORIUM The diagram at the right gives the dimensions of an auditorium. If new carpet is needed for the auditorium, what will be the area of the carpet? Round to the nearest square yard.

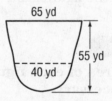
65 yd
55 yd
40 yd

SIDING For Exercises 10 and 11, use the diagram that shows one end of a cottage.

10. Each end of the cottage needs new siding. Find the total area that needs new siding.

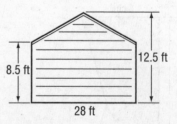
12.5 ft
8.5 ft
28 ft

11. The siding material costs $75 for a bundle of siding that covers an area of 100 square feet. What will be the total cost to put siding on both ends of the cottage? Justify your answer.

11-7 Study Guide and Intervention

Three-Dimensional Figures

Lesson 11-7

Prisms	At least 3 rectangular lateral faces	Top and bottom bases are parallel	Shape of the base tells the name of the prism
Pyramids	At least three triangular lateral faces	One base shaped like any 3-sided closed figure	Shape of the base tells the name of the pyramid
Cones	Only one base	Base is a circle	One vertex and no edges
Cylinders	Only two bases	Bases are circles	No vertices and no edges
Spheres	All points are the same distance from the center	No faces or bases	No edges or vertices

Example For each figure, name the shape of the base(s). Then classify each figure.

A.

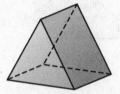

The figure has two parallel triangular bases and three rectangular faces. The figure is a triangular prism.

B.

The figure has two circular bases and no edges. The figure is a cylinder.

Exercises

For each figure name the shape of the base(s). Then classify each figure.

1.

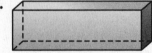

2.

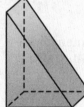

3.

4.

5.

6.

11-7 **Practice**

Three-Dimensional Figures

For each figure, identify the shape of the base(s), if any. Then classify the figure.

1.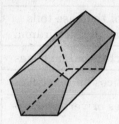

2.

3.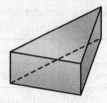

4.

5.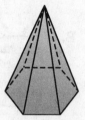

6.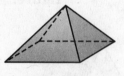

7.

8.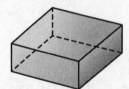

9.

10. **CANDLES** What three-dimensional figure describes the candle shown?

11. **FENCES** The basic shape of a fence post is made of two geometric figures. Classify these figures.

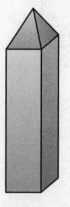

11-8 Study Guide and Intervention
Drawing Three-Dimensional Figures

A solid is a three-dimensional figure.

Example 1 Draw a top, a side, and a front view of the solid at the right.

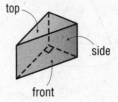

The top view is a triangle. The side and front views are rectangles.

top

side

front

Example 2 Draw the solid using the top, side, and front views shown below.

top

side

front

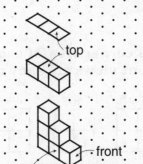

Step 1 Use the top view to draw the base of the figure, a 1-by-3 rectangle.

Step 2 Add edges to make the base a solid figure.

Step 3 Use the side and front views to complete the figure.

Exercises

1. Draw a top, a side, and front view of the solid.

2. Draw a corner view of the three-dimensional figure whose top, side, and front views are shown. Use isometric dot paper.

top   side   front

Lesson 11-8

11-8 **Practice**

Drawing Three-Dimensional Figures

Draw a top, a side, and a front view of each solid.

1.

2.

3.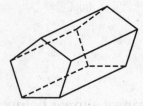

4.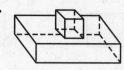

Draw a corner view of each three-dimensional figure whose top, side, and front views are shown. Use isometric dot paper.

1. top   side   front

2. top   side   front

7. **HAT RACK** Draw a top, a side, and a front view of the hat rack shown.

8. **MUSIC** Sketch views of the top, side, and front of the piano shown.

186

11-9 Study Guide and Intervention

Volume of Prisms

The **volume** of a three-dimensional figure is the measure of space occupied by it. It is measured in cubic units such as cubic centimeters (cm³) or cubic inches (in³). The volume of the figure at the right can be shown using cubes.

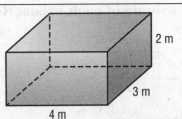

| The bottom layer, or base, has $4 \cdot 3$ or 12 cubes. | → | 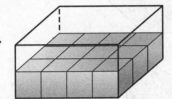 | 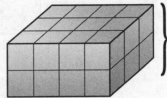 | There are two layers. |

It takes $12 \cdot 2$ or 24 cubes to fill the box. So, the volume of the box is 24 cubic meters.

A **rectangular prism** is a three-dimensional figure that has two parallel and congruent sides, or bases, that are rectangles. To find the volume of a rectangular prism, multiply the area of the base and the height, or find the product of the length ℓ, the width w, and the height h.

$V = Bh$ or $V = \ell wh$

Example   **Find the volume of the rectangular prism.**

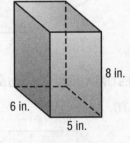

$V = \ell wh$ Volume of a rectangular prism

$V = 5 \cdot 6 \cdot 8$ Replace ℓ with 5, w with 6, and h with 8.

$V = 240$ Multiply.

The volume is 240 cubic inches.

Exercises

Find the volume of each rectangular prism. Round to the nearest tenth if necessary.

1.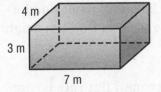

2.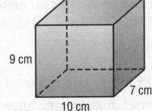

3.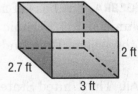

Lesson 11-9

11-9 Practice

Volume of Prisms

Find the volume of each prism. Round to the nearest tenth if necessary.

1.

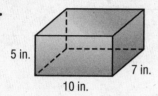

5 in.
7 in.
10 in.

2.

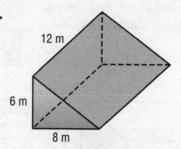

12 m
6 m
8 m

3.

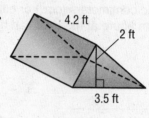

4.2 ft
2 ft
3.5 ft

4.

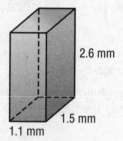

2.6 mm
1.5 mm
1.1 mm

5.

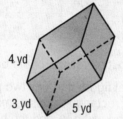

4 yd
3 yd
5 yd

6.

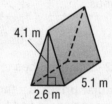

4.1 m
5.1 m
2.6 m

7.

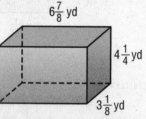

$7\frac{3}{4}$ yd
$5\frac{1}{2}$ yd
8 yd

8.

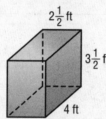

$2\frac{1}{2}$ ft
$3\frac{1}{2}$ ft
4 ft

9.

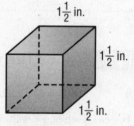

$1\frac{1}{2}$ in.
$1\frac{1}{2}$ in.
$1\frac{1}{2}$ in.

ESTIMATION Estimate to find the approximate volume of each prism.

10.

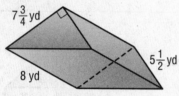

$6\frac{7}{8}$ yd
$4\frac{1}{4}$ yd
$3\frac{1}{8}$ yd

11.

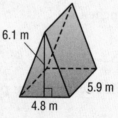

6.1 m
5.9 m
4.8 m

12. ALGEBRA The base of a rectangular prism has an area of 15.3 square inches and a volume of 185.13 cubic inches. Write an equation that can be used to find the height h of the prism. Then find the height of the prism.

13. MAIL The United States Post Office has two different priority mail flat rate boxes. Which box has the greater volume? Justify your answer. Box 1: $6\frac{1}{2}$ in. $\times$ $8\frac{1}{2}$ in. $\times$ 11 in. Box 2: $3\frac{3}{8}$ in. $\times$ $11\frac{7}{8}$ in. $\times$ $13\frac{5}{8}$ in.

11-10 Study Guide and Intervention

Volume of Cylinders

As with prisms, the area of the base of a **cylinder** tells the number of cubic units in one layer. The height tells how many layers there are in the cylinder. The volume V of a cylinder with radius r is the area of the base B times the height h.

$V = Bh$ or $V = \pi r^2 h$, where $B = \pi r^2$

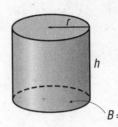

Example **Find the volume of the cylinder. Use 3.14 for π. Round to the nearest tenth.**

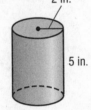

2 in.

5 in.

$V = \pi r^2 h$ Volume of a cylinder

$V \approx 3.14(2)^2(5)$ Replace π with 3.14, r with 2, and h with 5.

$V \approx 62.8$ Simplify.

The volume is approximately 62.8 cubic inches. Check by using estimation.

Exercises

Find the volume of each cylinder. Use 3.14 for π.
Round to the nearest tenth.

1.

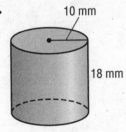

10 mm

18 mm

2.

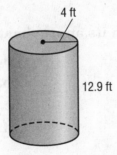

4 ft

12.9 ft

3.

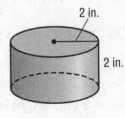

2 in.

2 in.

4. radius = 9.5 yd
   height = 2.2 yd

5. diameter = 6 cm
   height = 11 cm

6. diameter = $3\frac{2}{5}$ m

   height = $1\frac{1}{4}$ m

Lesson 11-10

11-10 Practice

Volume of Cylinders

Find the volume of each cylinder. Use 3.14 for π.
Round to the nearest tenth.

1.

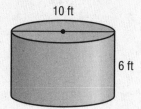
10 ft
6 ft

2.

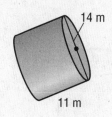
14 m
11 m

3.

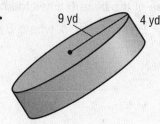
9 yd   4 yd

4.

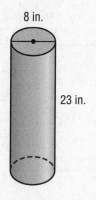
8 in.
23 in.

5.

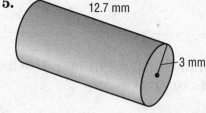
12.7 mm
3 mm

6.

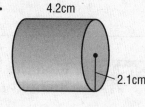
4.2cm
2.1cm

7. radius = 3.7 cm

height = 5.2 cm

8. diameter = 6 in.

height = $4\frac{1}{2}$ in

9. radius = $5\frac{1}{4}$ yd

height = $6\frac{1}{2}$ yd

10. CONTAINER What is the volume of a barrel that has a diameter of $1\frac{1}{2}$ feet and a height of 4 feet?

ESTIMATION Match each cylinder with its approximate volume.

11. diameter = 4 cm, height = 3.6 cm

12. radius = 2.7 cm, height = 5 cm

13. radius = 3 cm, height = 4.1 cm

14. diameter = 8.2 cm, height = 2 cm

a. 108 ft³

b. 135 ft³

c. 96 ft³

d. 48 ft³

15. FUEL Two fuel tanks with the dimensions shown have the same volume. What is the value of *h*?

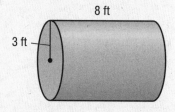
8 ft
3 ft

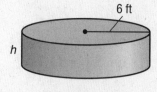
6 ft
h

12-1 Study Guide and Intervention

Estimating Square Roots

Recall that a perfect square is a square of a rational number. In Lesson 5-8, you learned that any number that can be written as a fraction is a rational number. A number that cannot be written as a fraction is an **irrational number**.

Example 1 Estimate $\sqrt{40}$ to the nearest whole number.

List some perfect squares.

1, 4, 9, 16, 25, 36, 49, …

$36 < 40 < 49$ 40 is between the perfect squares 36 and 49.

$\sqrt{36} < \sqrt{40} < \sqrt{49}$ Find the square root of each number.

$6 < \sqrt{40} < 7$ $\sqrt{36} = 6$ and $\sqrt{49} = 7$

So, $\sqrt{40}$ is between 6 and 7. Since 40 is closer to 36 than to 49, the best whole number estimate is 6.

Example 2 Graph $\sqrt{28}$ on a number line.

  28 [ENTER] 5.291502622

$\sqrt{28} \approx 5.3$

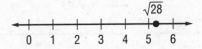

Check for Reasonableness Since $5^2 = 25$ and 25 is close to 28, the answer is reasonable.

Exercises

Estimate each square root to the nearest whole number.

1. $\sqrt{3}$ 2. $\sqrt{8}$

3. $\sqrt{26}$ 4. $\sqrt{41}$

5. $\sqrt{61}$ 6. $\sqrt{94}$

7. $\sqrt{152}$ 8. $\sqrt{850}$

Graph each square root on a number line.

9. $\sqrt{2}$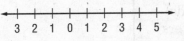

10. $\sqrt{27}$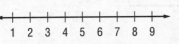

11. $\sqrt{73}$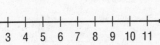

12. $\sqrt{82}$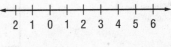

13. $\sqrt{105}$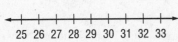

14. $\sqrt{395}$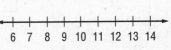

15. $\sqrt{846}$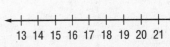

16. $\sqrt{2{,}298}$ (number line: 45 46 47 48 49 50 51 52 53)

Lesson 12-1

12-1 **Practice**

Estimating Square Roots

Estimate each square root to the nearest whole number.

1. $\sqrt{8}$ 2. $\sqrt{19}$ 3. $\sqrt{47}$ 4. $\sqrt{70}$

5. $\sqrt{91}$ 6. $\sqrt{125}$ 7. $\sqrt{150}$ 8. $\sqrt{389}$

9. $\sqrt{2,468}$ 10. $\sqrt{899}$ 11. $\sqrt{4,840}$ 12. $\sqrt{8,080}$

Graph each square root on a number line.

13. $\sqrt{6}$ 14. $\sqrt{21}$ 15. $\sqrt{53}$ 16. $\sqrt{79}$

3 2 1 0 1 2 3 4 5   2 1 0 1 2 3 4 5 6   1 2 3 4 5 6 7 8 9   6 7 8 9 10 11 12 13 14

17. $\sqrt{190}$ 18. $\sqrt{624}$ 19. $\sqrt{427}$ 20. $\sqrt{3,178}$

7 8 9 10 11 12 13 14 15   20 21 22 23 24 25 26 27 28   18 19 20 21 22 23 24 25 26   51 52 53 54 55 56 57 58 59

21. $\sqrt{0.36}$ 22. $\sqrt{0.81}$ 23. $\sqrt{1.44}$ 24. $\sqrt{2.25}$

3 2 1 0 1 2 3 4 5   2 1 0 1 2 3 4 5 6   1 2 3 4 5 6 7 8 9   1 0 1 2 3 4 5 6 7

25. **ALGEBRA** What whole number is closest to $\sqrt{a + b}$ if $a = 24$ and $b = 38$?

26. **ALGEBRA** Evaluate $\sqrt{x - y}$ to the nearest tenth if $x = 10$ and $y = 4.5$

27. **QUILTING** A queen-size quilt in the shape of a square has an area of 51 square feet. What is the approximate length of one side of the quilt to the nearest tenth?

28. **PENDULUM** The formula below can be used to estimate the time it takes for a pendulum to swing back and forth once. Use the formula to find the time it takes for a pendulum with a length of 0.8 meter to swing back and forth once. Round to the nearest tenth.

$$T = 2 \times \sqrt{L}$$
- T = time (seconds)
- L = length (meters)

12-2 Study Guide and Intervention

The Pythagorean Theorem

The sides of a right triangle have special names. The sides adjacent to the right angle are the **legs**. The side opposite the right angle is the **hypotenuse**. The **Pythagorean Theorem** describes the relationship between the length of the hypotenuse and the lengths of the legs. In a right triangle, the square of the length of the hypotenuse equals the sum of the squares of the lengths of the legs.

$$c^2 = a^2 + b^2$$

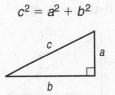

Example 1 Find the missing measure of a right triangle if $a = 4$ inches and $b = 3$ inches.

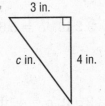

$c^2 = a^2 + b^2$	Pythagorean Theorem
$c^2 = 4^2 + 3^2$	Replace a with 4 and b with 3.
$c^2 = 16 + 9$	Evaluate 4^2 and 3^2.
$c^2 = 25$	Add.
$\sqrt{c^2} = \sqrt{25}$	Take the positive square root of each side.
$c = 5$	Simplify.

The length of the hypotenuse is 5 inches.

Example 2 Determine whether a triangle with side lengths of 6 meters, 9 meters, and 12 meters is a right triangle.

$c^2 = a^2 + b^2$	Pythagorean Theorem
$12^2 \overset{?}{=} 6^2 + 9^2$	Replace a with 6, b with 9, and c with 12.
$144 \overset{?}{=} 36 + 81$	Simplify.
$144 \neq 117$	Add.

The triangle is *not* a right triangle.

Exercises

Find the missing measure of each right triangle. Round to the nearest tenth if necessary.

1.

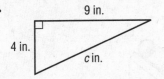

2.

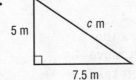

3.

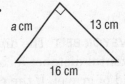

Determine whether each triangle with the given side lengths is a right triangle. Write *yes* or *no*.

4. 15 ft, 8 ft, 17 ft   **5.** 5 in., 13 in., 17 in.   **6.** 9 yd, 40 yd, 41 yd

12-2 Practice

The Pythagorean Theorem

Find the missing measure of each triangle. Round to the nearest tenth if necessary.

1.
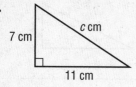
7 cm c cm   11 cm

2.
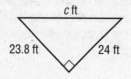
c ft   23.8 ft   24 ft

3.
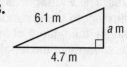
6.1 m a m   4.7 m

4. $a = 3.3$ in., $b = 5.6$ in.

5. $b = 2.9$ mm, $c = 4.4$ mm

6. $a = 21$ yd, $c = 29$ yd

7. $a = 2\frac{1}{5}$ ft, $c = 4\frac{2}{5}$ ft

8. $b = 7\frac{1}{4}$ in., $c = 7\frac{3}{4}$ in.

9. $a = 6\frac{1}{2}$ yd, $b = 10$ yd

If a triangle has sides a, b, and c so that $a^2 + b^2 = c^2$, then the triangle is a right triangle. Determine whether a triangle with the given side lengths is a right triangle. Write *yes* or *no*.

10. 9 cm, 12 cm, 18 cm

11. 7 ft, 24 ft, 25 ft

12. 5 in., 12 in., 13 in.

Find the missing measure in each figure. Round to the nearest tenth if necessary.

13.
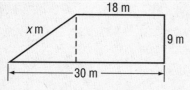
18 m x m   9 m   30 m

14.
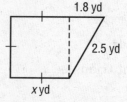
1.8 yd   2.5 yd x yd

15. **SOCCER** Find the width of the soccer goal. Round to the nearest tenth.
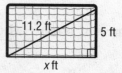
11.2 ft   5 ft x ft

16. **CONVEYOR BELT** The diagram shows the horizontal distance a conveyor belt moves a load of gravel. If the conveyor belt takes 54 seconds to move gravel from the bottom of the conveyor belt to the top at a rate of 3 feet per second, how high does the conveyor belt lift the gravel? Round to the nearest tenth.
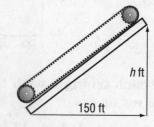
h ft   150 ft

12-3 Study Guide and Intervention

Problem-Solving Investigation: Make a Model

When solving problems, make a model to represent the given situation in order to determine the best plan for a solution.

Example GIFT WRAP Rita wants to wrap a rectangular box. The box is 12 inches by 7 inches by 3 inches high. What must be the area of the paper so that she has a 1 inch overlap to neatly wrap the paper?

Understand You know that the box is $12 \times 7 \times 3$ and that you need to add 1 inch to some measures for the overlap. You also know that the wrapping paper will be a rectangle.

Plan Draw a sketch of the box and then make a model of the box if it were cut apart and laid flat. You need the overlap going around the box.

Solve Sketch the box.   Make a model of the box unfolded.

The length of the paper needed is the distance around the box plus 1 inch. So, $\ell = 7 + 3 + 7 + 3 + 1$ or 21 inches.

The width of the paper would be $3 + 12 + 3$ or 18 inches.

The area would be 21×18 or 378 in^2.

Check Make a box using centimeters instead of inches. Then cut a piece of paper 18 centimeters by 21 centimeters to see if you can wrap the box neatly.

Exercises

1. **GARDENING** Peg wants to put a stone path 3 feet wide around her rectangular garden measuring 10 feet by 15 feet. What will be the perimeter of her garden including the stone path?

2. **DRAWING** Dante is making a full-size drawing of his favorite cartoon character. If the figure is 1 inch by 0.5 inches and his scale is 1 inch = 10 inches, how large will the full size character be?

12-3 Practice

Problem-Solving Investigation: Make a Model

Mixed Problem Solving

For Exercises 1 and 2, make a model to solve the problem.

1. **ARCHITECT** Mrs. Peron is designing a home for a client. The house is 45 feet by 76 feet. If she uses a scale of 1 foot = $\frac{1}{2}$ inch, what are the dimensions of the house on the blue prints?

2. **SWIMMING POOL** Mr. Forrester has a swimming pool that measures $3\frac{1}{3}$ yards by 8 yards. If the deck around the pool is $2\frac{2}{3}$ yards wide, what is the outside perimeter of the deck?

Use any strategy to solve Exercises 3 through 6. Some strategies are shown below.

PROBLEM-SOLVING STRATEGIES
• Draw a diagram.
• Use logical reasoning.
• Make a model.

3. **BATTERIES** A manufacturing plant can make 350 batteries in 15 minutes. How long will it take the manufacturing plant to make 3,500 batteries?

4. **SHOPPING** A grocery store has five cash registers. About 4 customers are checked out at each register every 20 minutes. How many customers are checked out at the store each hour?

5. **TESTS** Diego scored a 95 on his first test in science class. He then scored 100 on his next 5 tests. If he scored a 91 on his seventh test, what is his test average?

6. **NEWSPAPERS** Candace wants to increase the number of newspapers she delivers. She currently delivers 58 newspapers. In fourteen weeks, she wants to be delivering 100 newspapers. How many newspaper deliveries must she increase each week to obtain her goal?

12-4 Study Guide and Intervention

Surface Area of Rectangular Prisms

> The sum of the areas of all the surfaces, or faces, of a three-dimensional figure is the **surface area**. The surface area S of a rectangular prism with length ℓ, width w, and height h is found using the following formula.
>
> $S = 2\ell w + 2\ell h + 2wh$

Example Find the surface area of the rectangular prism.

You can use the net of the rectangular prism to find its surface area. There are three pairs of congruent faces in a rectangular prism:

- top and bottom
- front and back
- two sides

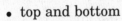

Faces	Area
top and bottom	$(4 \cdot 3) + (4 \cdot 3) = 24$
front and back	$(4 \cdot 2) + (4 \cdot 2) = 16$
two sides	$(2 \cdot 3) + (2 \cdot 3) = 12$
Sum of the areas	$24 + 16 + 12 = 52$

Alternatively, replace ℓ with 4, w with 3, and h with 2 in the formula for surface area.

$S = 2\ell w + 2\ell h + 2wh$

$\quad = 2 \cdot 4 \cdot 3 + 2 \cdot 4 \cdot 2 + 2 \cdot 3 \cdot 2$ Follow order of operations.

$\quad = 24 + 16 + 12$

$\quad = 52$

So, the surface area of the rectangular prism is 52 square meters.

Exercises

Find the surface area of each rectangular prism.

1.
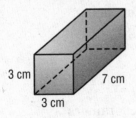

2.
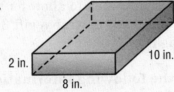

3.
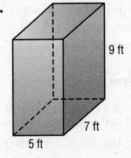

Lesson 12-4

12-4 Practice

Surface Area of Rectangular Prisms

Find the surface area of each rectangular prism. Round to the nearest tenth if necessary.

1.

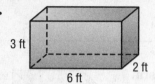

3 ft
6 ft
2 ft

2.

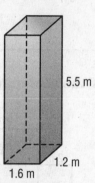

5.5 m
1.6 m
1.2 m

3.

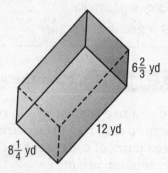

$6\frac{2}{3}$ yd
12 yd
$8\frac{1}{4}$ yd

4. length = 20 cm
width = 18 cm
height = 25 cm

5. length = 31.5 in.
width = 12.2 in.
height = 24.8 in.

6. length = 5.3 mm
width = 1.1 mm
height = 3.4 mm

7.

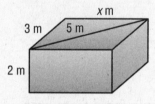

x m
3 m   5 m
2 m

8.

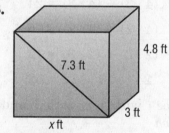

4.8 ft
7.3 ft
3 ft
x ft

ESTIMATION Estimate the surface area of each prism.

9.

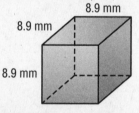

8.9 mm
8.9 mm
8.9 mm

10.

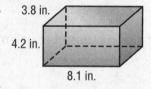

3.8 in.
4.2 in.
8.1 in.

11. BIRTHDAY GIFT When wrapping a birthday gift for his mother, Kenji adds an additional 2.5 square feet of gift wrap to allow for overlap. How many square feet of gift wrap will Kenji use to wrap a gift 3.5 feet long, 18 inches wide, and 2 feet high?

For Exercises 12 and 13, use the following information.

A company needs to package hazardous chemicals in special plastic containers that hold 80 cubic feet of chemicals.

12. Find the whole number dimensions of the container that would use the least amount of plastic.

13. If the plastic costs $0.10 per square foot, how much would it cost to make 24 containers?

12-5 Study Guide and Intervention

Surface Area of Cylinders

The diagram below shows how you can put two circles and a rectangle together to make a cylinder.

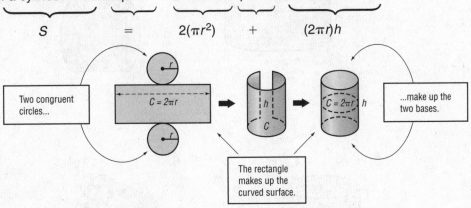

The surface area of a cylinder equals the area of two bases plus the area of the curved surface.

$$S = 2(\pi r^2) + (2\pi r)h$$

Two congruent circles...

The rectangle makes up the curved surface.

...make up the two bases.

In the diagram above, the length of the rectangle is the same as the circumference of the circle. Also, the width of the rectangle is the same as the height of the cylinder.

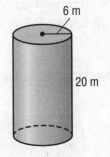

6 m
20 m

Example Find the surface area of the cylinder. Use 3.14 for π. Round to the nearest tenth.

$S = 2\pi r^2 + 2\pi rh$ Surface area of a cylinder.

$S = 2\pi(6)^2 + 2\pi(6)(20)$ Replace π with 3.14, r with 6, and h with 20.

≈ 979.7 Simplify.

The surface area is about 979.7 square meters.

6 m
20 m

Lesson 12-5

Exercises

Find the surface area of each cylinder. Use 3.14 for π. Round to the nearest tenth.

1.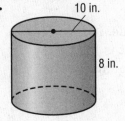
10 in.
8 in.

2.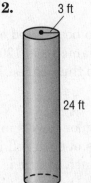
3 ft
24 ft

3.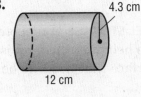
4.3 cm
12 cm

12-5 Practice

Surface Area of Cylinders

**Find the surface area of each cylinder. Use 3.14 for π.
Round to the nearest tenth.**

1.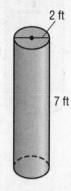
2 ft
7 ft

2.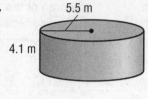
5.5 m
4.1 m

3.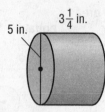
$3\frac{1}{4}$ in.
5 in.

4.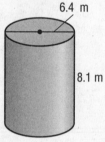
6.4 m
8.1 m

5.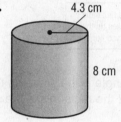
4.3 cm
8 cm

6.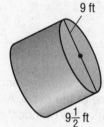
9 ft
$9\frac{1}{2}$ ft

7. diameter = 15.2 mm
height = 9.4 mm

8. diameter = 28.4 yd
height = 15.1 yd

9. radius = 50 cm
height = 70 cm

ESTIMATION Estimate the area of each cylinder.

10.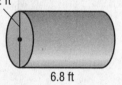
4.2 ft
6.8 ft

11.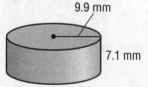
9.9 mm
7.1 mm

12.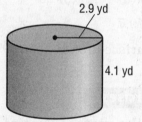
2.9 yd
4.1 yd

13. FUEL STORAGE A fuel storage tank needs to be painted on the inside. If the height of the tank is 40 feet and the diameter is 120 feet, what is the surface that needs to be painted? Round to the nearest hundred square feet.

14. PAPER TOWELS Each of the three rolls of paper towels in a package are individually wrapped in plastic. The radius of each roll is 5.6 centimeters and the height is 27.9 centimeters. How much plastic is used to individually wrap the three rolls? Round to the nearest tenth.